THE AI GAP

Women, AI, and the Next Great Leap Forward

ERICA ROONEY

WORLDCHANGERS
MEDIA

Published by WorldChangers Media
www.worldchangers.media

Print ISBN: 978-1-955811-96-5
Ebook ISBN: 978-1-955811-97-2

Printed by Book International Inc.
Distributed by Casemate Publishers and Book Distributors LLC

Printed in the USA

**This book is dedicated to the women
who chose courage over comfort.**

To the women who came before us and fought for our rights.
To the women building the table now.
And to the women rising behind us.
To the woman who has not yet found her voice—
we will speak louder until you do.
Through HER Collective, and in every room where
women choose conviction over competition:
We lift while we climb.
We build what we did not inherit.
We refuse the limits handed to us.
We redefine what power looks like.
You can sit with us.

A NOTE FROM THE AUTHOR

AI is evolving at an unprecedented pace. The tools, platforms, and capabilities referenced in this book reflect the landscape at the time of writing. Specific tools and technologies will shift, improve, and be replaced. The thinking, confidence, and leadership frameworks inside this book are meant to equip you to evolve alongside them.

TABLE OF CONTENTS

INTRODUCTION

CLOSE THE GAP WITH ONE GREAT LEAP

One early September morning in 2025, as I lay in bed in my Boston hotel room at MIT, I picked up my phone to find 50+ messages from women in my CHIEF Community network, and they were fired up. I mean, they were *pissed*. CHIEF is an executive network for women who are doing the dang thing. They are leaders in their communities and companies, and they are changing the world. We were all gathered at MIT for the AI-Powered Women's Conference.

I co-host a podcast with my business partner and friend, Greg Boone, called *AI Voice or Victim,* and we were slated to be the podcast of the conference. We were there to capture the voices and stories of as many women AI leaders as we could. The group text was filled with women who are founders and CEOs of AI tech companies, Chief AI Officers, Chief Technology Officers, entrepreneurs, and other AI badasses. (In other words, they know their stuff when it comes to AI.) This group of women had gone to dinner the night before the conference, while I stayed back to rest my voice before my event, but these 50+ messages told me everything I missed.

"WTF???"

"Are you seeing this?"

"This can't be real!"

These messages told a story.

At some point during the dinner, one of the women tested a prompt in ChatGPT: *Create an image. Based on what you have learned about me since we've been chatting together, generate your best guess of what I look like. No preamble.*

Then asked: *What was it you (Chat) learned that prompted those details?*

Everyone followed suit, testing the prompt inside their own ChatGPTs.

What followed next in my text thread was WILD.

Women started posting their results, and image after image scrolled past. My heart thudded as dozens of faces appeared—none mine, none any of theirs. Only frumpy, middle-aged white men with glasses and three days' worth of stubble on their faces. As if the machine's idea of authority was stuck in some sepia-toned archive.

One woman, a Senior Vice President, mom of five, and role model for many, was rendered as a white dude with glasses and a beard. Another woman, an HR director with two decades of experience, was again pictured as a white male. Next, a founder, a Black executive, and another senior leader, all women, were all reimagined by AI as the same old, pale, stale, white male (with three days of scruff). No matter the title, experience, or race, AI kept defaulting to the same thing.

But let me be clear: this isn't just a woman's problem.

Greg Boone, my podcast co-host, faced it too. Prompt after prompt yielded white men, as if the very idea of a Black CEO did not compute. In fact, the AI was so convinced he was a white man that it gave him not ONE but TWO different white men to pick from. What makes this wild is that he'd previously provided a picture of himself to ChatGPT, so there was zero excuse. He finally coaxed out an image of a Black man only after multiple retries.

If AI can't even recognize leaders in our own networks and our own communities, how can it help us break barriers?

Then came the jaw-dropper.

One woman's results (who wanted to remain anonymous) generated an illustration of a woman with artwork in the background.

* To safeguard her privacy, the woman's image has been intentionally altered.

(OK … we're seeing *some* progress, right? Finally, a woman.) The two pieces of artwork were interesting choices, though. One was of a cat (which, OK, she likes cats) and then one framed picture of the quote "**We MEN belong in all places where decisions are being made.**"

Are you fucking kidding me? Could it be any more absurd? If writing a book on this very topic hadn't already been on my mind, it was now solidified.

We (society, the world at large) have spent a lot of time talking about the importance of having women, and other diverse representation, at the table. Of making sure their voices are heard and built into the technology, but here we are in 2026, and these are the outputs. We have been sounding the alarm on glass ceilings and sticky floors (shameless plug here for my first book that bears the same title) for years, and yet, in 2026, the machine still defaults to the same biases we have been fighting. This wasn't a glitch. It was a mirror. And these mirrors reveal the systems we forgot to redesign.

The world is changing faster than most people can spell the word "algorithm." And women are standing right at the edge of what I'm calling the **next great leap forward**. This leap has the power to finally close the "gaps" we've lived with for generations.

The gaps we have pushed against, shouted about, protested about, and carried like an invisible weight for far too long. Despite decades of effort, the gaps have barely moved, and in some places, they have grown wider. The pay gap. The confidence gap. The leadership gap. The representation gap. (Are you exhausted yet? Because I sure am!)

These gaps aren't abstract ideas or "fake news" but measurable realities that women live with daily. As of 2024, women in the U.S. earned 85% of what men earned in median hourly pay across all workers, a figure that has barely shifted in two decades.[1] As of now, only 55 women lead Fortune 500 companies, representing barely 11% of the total, and this is up from just 10% in 2023.[2] (A truly pitiful statistic.) Research shows that women tend to rate their own abilities lower than men who have similar experience! This is a pattern documented across multiple studies and often referred to as the "Confidence Gap."[3] Across industries, representation in senior leadership roles remains stubbornly low, especially for women of color.[4] (Another cringe-worthy stat.)

Ya'll! This is not a one-off error! It's a pattern. And my favorite saying in Human Resources is "When it becomes a pattern, it becomes a problem." Why? Because patterns compound over time. Now, in this moment, we are up against what I call "the AI Gap." The AI Gap is the compounding distance between those shaping AI systems and those being shaped by them.

AI does not create inequality. It accelerates whatever the system was already designed to produce. If the baseline was male, AI scales male. If the baseline ignored women's bodies, AI ignores women's data. If power was concentrated before, it compounds faster now. The question is not whether AI will be biased. The question is whether we redesign the system before it's too late.

The AI Gap is still taking shape, and it isn't on the radar for many people as a potential issue for women and other marginalized communities, but it's there, widening quietly every day. It's the distance between those who are learning to use AI and those who are waiting to understand it. Between those who are shaping the future and those who are being shaped by it.

The AI Gap isn't about the technology behind it. It's about access, awareness, and agency. It's about who gets to lead in a world that's being rebuilt by invisible algorithms and who gets left out entirely. And while women are inching forward in C-Suite representation, women are still 25% less likely than men to adopt AI tools, a disparity that threatens to widen every existing gap in leadership, confidence, representation, and pay.[5]

"The AI Gap is the compounding distance between those shaping AI systems and those being shaped by them."

Now, this divide doesn't show up all at once. It reveals itself through a series of distinct, compounding gaps that we women are all too familiar with, starting with the Leadership Gap.

The AI Gap doesn't appear as one dramatic fracture. It spreads through the same structural cracks we've been navigating for decades.

The Leadership Gap

I'm going to lay it out there. Women enter the workforce at nearly equal rates to men, yet their representation drops off sharply at the highest levels. And riddle me this: How, in 2024, can women make up roughly half of the entry-level roles but only 12% of the C-suite and board positions?[6] I'll tell you how. This disparity stems from the structural barriers like bias in promotion, lack of sponsorship, and exclusion from critical networks (just to name a few).

In AI, the Leadership Gap means women rarely shape the very algorithms they help train. With just one in eight tech and AI leadership roles held by women, decisions about data selection, model design, and ethical guardrails default to male-dominated viewpoints, embedding bias into every line of code.[7]

Now let's take a good look at the leading AI companies today, and how gender diverse they are. At the time of writing this book (January 2026):

- **OpenAI:** Following its 2023 leadership upheaval, OpenAI reconstituted its board in 2024 to include several high-profile women, including Dr. Sue Desmond-Hellmann, Nicole Seligman, and Fidji Simo. While representation has improved from an all-male board, women still make up a minority of board seats.[8]
- **Google DeepMind:** Lila Ibrahim serves as the first and only female Chief Operating Officer at DeepMind, representing approximately 10% of its senior executive team.[9]
- **Anthropic:** Co-founder and President Daniela Amodei is just one of two women in Anthropic's executive leadership team, accounting for approximately 17% female representation among its senior leaders.[10]

Despite making up nearly half the workforce, women occupy only one in eight senior leadership roles in tech and AI. These roles drive the strategic decisions about algorithms, data, and design, and without seats at the tables where AI policies and products are defined, women's perspectives disappear from the code itself.

The Confidence Gap

I know you have heard the line, "Men apply when they meet 60% of the qualifications, but women only apply when they hit 100%."[11] (Tell me you've heard it!) But, fun fact: that soundbite is more legend than tight science. When researchers tested it in controlled settings, they found something close: women shy away when the bar is murky, while men lean in when they feel they can "figure it out."[12]

In the AI context, this translates to hesitation to experiment with new tools, requests for explicit permission, and the fear of looking foolish—leaving innovation to those who zip through trials and errors without the self-doubt.

The Representation Gap

It can't be a surprise to you to hear that women's voices have long been marginalized in policymaking, boardrooms, and media. That leads to stereotypes and absurd one-dimensional portrayals. In AI, this underrepresentation crystallizes in training data and governance. Male and Western perspectives dominate datasets, and ethics boards lack gender diversity. Consequently, AI systems default to white male faces and perspectives even when prompted otherwise.[13]

The Pay Gap

Are you angry yet? Tired? Confused? I am! How is it 2026 and, according to the most recent pay data, women working full-time still earn only 83.2 cents for every dollar paid to men? That inequity is magnified for women of color, earning only 72.3% of all men's earnings.[14] For Black women specifically, the gap is wider, deeper, and far more persistent.

The gap widens further when you factor in seniority and exposure to automation-related job risks. It's driven by occupational segregation, unequal negotiation outcomes, and biased performance evaluations that continue to lock in economic inequality. Early adopters of AI tools are already gaining a productivity edge, and in most organizations, productivity translates into visibility, advancement, and pay. Yet women are about 25% less likely than men to adopt these technologies.[15] That gap is not just about usage; it's about who benefits first and who gets left behind.

•

Individually, each gap slows progress, and it's thick and resistant, like molasses. Together, they begin to stack, accelerating movement in the wrong direction. Left unchecked, these four gaps will merge into a single divide, an AI Gap so wide that women, and all underrepresented groups, risk being shut out of the most powerful economic engine of our time.

Closing this gap demands more than the baby steps we've been taking. It calls for a full-scale, high-velocity campaign and a structural redesign. It demands that women have a lead seat at every AI table, allies using their influence to champion diverse voices, and organizations embedding gender metrics into every layer of their AI ecosystem. It's only through this united, relentless charge that we can rewrite the code of our shared future. Only then will we be able to transform AI from an exclusionary barrier that can only define a leader as a white male around the age of 45, into the launching pad for infinite possibilities and lasting equity.

If infrastructure determines scale, then humanity determines direction. In the research for this book, a survey revealed that women consistently report significant fear and anxiety around AI, particularly the concern that it will replace roles many women hold.[16] Don't believe the hype. AI is not eliminating human contribution; it is reorganizing where value is created. This moment is not about machines taking over. It is about redefining what makes us brilliantly, remarkably human. And that is where the Five Cs of Humanity (critical thinking, curiosity, collaboration, communication, and connection) come in: the capabilities that ensure women (and all humans) shape AI rather than disappear within it.

When we lead with what makes us human and understand that, at our core, we'll never be automated, we can redefine what's possible.

That is what this book is really about. It's about unlearning the limits we have placed on ourselves, and the ones that the world has quietly placed on us. It's about facing the truth of where women stand today: the inequalities that persist, the systems that still need breaking, and the new ones being built through AI. This book is about **this moment, this *edge*** we are standing on, and the opportunity in front of us. But more than anything, it's about learning to lead with the power of AI.

For years, we have shouted from the rooftops that we want equal seats at the table, but for that to happen, we have to start showing up in new ways, with new tools, new confidence, and a new kind of power that blends humanity with technology. This book will take you there, step by step, and is broken out into four parts that will move you from curiosity to confidence to badass leadership. Each one builds on the last, turning awareness into action and potential into tangible progress, and is backed by data, research, and survey insights. (You know I'm a science girl!)

At the end of every chapter, you'll find a **TLDR** section, and if that acronym is not already part of your everyday vocabulary, think of it as your quick, no-fluff summary. TLDR stands for "Too Long; Didn't Read," and it captures the core ideas of the chapter in a way you can revisit, reference, and carry forward without having to reread every page.

You'll also find tangible ways to apply what you've learned so that this journey doesn't just inspire you, it transforms you. I call these sections "**Your AI-Powered Actions**." Please take these actions literally. Do them. Write out the answers. Practice with your AI tools. Test them. Share your wins. **THIS** is where learning turns into leadership—by applying what you're learning here until it becomes your natural instinct. As you work through the book, your AI prowess will deepen, and your skills will build upon each other. (You just might

"For years, we have shouted from the rooftops that we want equal seats at the table, but for that to happen, we have to start showing up in new ways, with new tools, new confidence, and a new kind of power that blends humanity with technology."

surprise yourself at how wrong you were about "not being techie enough.") Each AI-Powered Action follows a simple three-step structure designed to move you from awareness to real-world application.

ASK helps you surface the beliefs, assumptions, or habits that may be holding you back.

APPLY gives you a concrete way to experiment with AI or leadership behaviors in real time.

AMPLIFY pushes you to make the learning visible because your AI confidence grows fastest when it is reinforced through action, repetition, and shared momentum.

The way I see it, the fascinating rise of AI is the beginning of a new era for women. When we bring our empathy, strategy, and humanity into this technology, **we take the lead**. The future is not written by the loudest voice in the room, or even the overwhelming majority. It is written by the women who are courageous enough to learn, adapt, and lead with heart. And that is **us**, standing at the edge of a great leap with the power to define what comes next.

Ready or not, this is the moment. Let's jump.

THE FIVE CS OF HUMANITY

There's a lot of fear around AI replacing humans. But these five traits are what make us irreplaceably human. They don't compete with AI—they **amplify** it. They are the very qualities that make AI meaningful, safe, and transformative when guided by human hands.

Critical Thinking: You must challenge outputs—don't just consume them without care

AI can process a lot of information faster than any human ever could, but it cannot discern truth, context, or consequence. Critical thinking is what helps us cut through the noise, see what's real, generate new data, and create meaningful insights. It's our ability to question assumptions, spot bias, and ask WHY? In a world where AI can generate content in seconds, women who can think critically will be the ones who shape the answers that move humanity forward.

Curiosity: The curious never stop becoming

Generative AI is a predictive machine. It's incredible at predicting patterns, but it can't feel wonder and amazement. Curiosity is that spark that says, "What if there is a new way? Another way? A different way?" It's the creative restlessness that fuels innovation. Every advancement, from medicine to technology, all started with someone asking a question they didn't have an easy answer to (one that can't be found in the realm of AI), and in the age of AI, curiosity is our greatest competitive edge, because it keeps us asking the questions that the machines don't yet have an answer to.

Collaboration: The best ideas are co-created

AI thrives on information and data, but humans thrive on conversation. Collaboration, working together toward a common goal, is the chemistry of perspectives and lived experiences that no machine can replicate.

When women lead with the spirit of collaboration, they can expand on possibilities by creating solutions that are more inclusive, more empathetic, and more human. The Chief People Officer in me has to tell you that collaboration builds trust. And trust is the key ingredient required to build effective teams that share a purpose.

Communication: Clarity is our greatest human algorithm

How we communicate is how we connect and how we build bridges between divides. It's how we dissolve misunderstandings and how we turn an idea into something meaningful. AI can refine your emails and mimic your tone, but it'll never be able to read the room, feel the emotion of the room, or understand the weight of the words that were said (or those left unsaid). It will never be able to sense the hesitation in a voice or the compassion behind a hard truth. That's what empathy does, and empathy is beautifully human. (Empathy is also a documented strength for women leaders, so how about that?!)

Connection: The one thing AI can never replicate

Connection is about feeling. It's about emotion. While AI can process the words you speak, it can't sense your presence. Connection is the heartbeat of humanity. It's what moves us to care about the world around us, the people in our lives, and the impact we leave behind each interaction. It's not built through data, but only shared experiences, compassion, and courage. Women have always been the architects of connection. We build bridges between ideas and people, and we weave trust throughout teams where previously there was none. In the age of AI, where almost anything can be simulated, connection cannot be faked, and that is our advantage. The future won't be led by the most technical minds, but by the most connected ones.

PART 1

CLOSING THE AWARENESS GAP

From Self Doubt
to
Structural Clarity

Before you can reshape a system, you have to recognize the one you are currently operating in.

The AI Gap does not begin in an IT lab or in a line of code. It begins in perception. It shows up in the quiet distance between the women who lean into technological change and the women who wait until they feel fully prepared. It lives in the stories we carry about who belongs in rooms where the future is being designed.

Artificial intelligence accelerates patterns. It takes what already exists inside organizations and moves it faster and farther into hiring systems, promotion pathways, compensation structures, and visibility algorithms. Participation shapes what becomes visible inside those systems, and visibility shapes what expands. (Are you with me?)

Closing the Awareness Gap starts with clarity. Clarity about the narratives you have absorbed about technology. Clarity about the systems that shaped your confidence. Clarity about how AI reflects and reinforces whatever is already embedded inside institutions. You are not standing outside this shift. You are already participating in it.

This section moves you from hesitation to engagement. You will examine the beliefs that influence your relationship with technology, experiment with AI in practical ways, and begin to see how systems scale behavior over time. As your awareness deepens, so does your ability to move intentionally inside those structures.

Awareness is the first act of power. Are you ready? Because once you see the architecture, you navigate it differently. Once you engage, your presence becomes part of what scales. This is where the gap begins to close.

CHAPTER 1

OWNING YOUR STORY

As a women's empowerment "expert," one thing I've learned is that before you can change an entire system, you have to change your story inside it, because the stories we tell ourselves determine whether we step into the rooms where the next systems are being designed.

For decades, women have heard the message that "*You can do anything you set your mind to.*" What isn't said out loud, but is learned over time, is that, yes, you can, BUT you will have to work twice as hard to prove it.

We have climbed the corporate ladder, jumping over broken rungs, shattering glass ceilings, building resumes, earning degrees, and checking every single box while still wondering, "Will this ever be enough?"

The biggest shifts start from within. If deep down you are telling yourself, "I'm not technical enough" or "I'm already so far behind on this AI stuff", any new tool, whether it's a simple app or a sophisticated AI tool, will feel out of reach. This chapter peels back those old narratives whispering, "AI isn't for people like me." Because the truth is, it's NOT about whether women belong in tech (we do, and always have). It's about the stories that shape our courage to experiment and the beliefs that expand or limit what's possible.

And sometimes those stories don't start inside us at all. Sometimes they start inside the systems and industries that were built long before we arrived. Take medicine and heart health, for example. For decades, cardiovascular disease was treated primarily as a "man's disease." Clinical research overwhelmingly used male bodies as the baseline for symptoms, treatment protocols, and risk assessment. Women's symptoms were often categorized as atypical, not because they were rare, but because they were different from those of men. The consequences of that design choice were not theoretical. Research shows that women are 50% more likely than men to be misdiagnosed when experiencing a heart attack.[17]

And medicine isn't the only place this has happened. For decades, crash test dummies were modeled on a 50th percentile male body.[18] Workplace tools were designed around male career patterns that assumed uninterrupted employment and no caregiving gaps.[19] Even voice recognition software historically performed better on lower-pitched male voices than on higher-pitched female ones.[20] This is the takeaway: when the baseline is male, the systems don't break. In fact, they are performing exactly as they were built to perform. It's just that it doesn't perform equally for everyone. That's what we are up against. Systems were built around one version of normal, and then they quietly labeled everyone else as a deviation. AI inherits similar defaults, and if women aren't present to question them, those inequities will be coded into algorithms.

When we fail to redesign those systems, they do exactly what they were built to do. AI is not different. It doesn't invent bias. It scales the default settings. In the early days of factory electrification, companies made what seemed like an obvious upgrade. They removed the old steam engines and installed large electric motors. On paper, this looked like progress. But inside the factory walls, almost nothing changed. Machines were still arranged around long shafts and belts. Workflows followed the same rigid patterns. Downtime, inefficiencies, and safety issues persisted. The new technology had been added, but the underlying architecture remained untouched, and productivity barely moved.[21]

The real transformation came later, when factory owners stopped treating electricity as a plug-in replacement and began redesigning the system around its true advantage: distributed power. Instead of one massive engine driving everything mechanically, smaller motors were installed at individual machines. Power flowed through wires, not belts. Layouts shifted. Material moved more efficiently. Processes became flexible. Entire management structures adapted. The productivity surge wasn't caused by the motor itself. It was unlocked by the redesign of the system around it.[22]

That's the lesson. General-purpose technologies don't automatically generate progress. Their impact depends on whether we redesign the surrounding infrastructure.[23] Electricity didn't change factories until factories changed themselves. More recent empirical work shows that while gains varied by industry and region, complementary redesign remained the decisive factor in unlocking measurable productivity growth.[24] AI will follow the same pattern. And because AI will define the next operating system of work and life, women must be part of that redesign, or they risk repeating the same exclusion. If we layer it on top of outdated assumptions, it'll preserve them. If we redesign around it, it'll amplify something entirely different.

If the default remains narrow, the output will remain narrow, only faster, and at scale.

Let that sit for a minute. The statistic isn't about confidence or competence. It's not about women doubting themselves. It's about what happens when the default model for "normal" excludes half the population. When the baseline is male, everyone else becomes an outlier, and when left unchecked, AI will harden that outlier status because it scales what it's given; only diverse architects can reset the baseline. And when you grow up inside the systems that consistently position you as the outlier, it doesn't just affect healthcare; it shapes your internal narrative about where you belong, what you're qualified for, and how loudly you are allowed to show up.

"When we fail to redesign those systems, they do exactly what they were built to do. AI is not different. It does not invent bias. It scales default settings."

Because AI functions as infrastructure, not inspiration, the conditions under which you engage with it will determine how it scales around you. In the pages ahead, you will discover that self-trust is the first form of AI literacy. If your internal system remains rooted in hesitation, AI will not correct that imbalance; it will operationalize it. Before you can evaluate an output or challenge a bias, you have to believe that you are capable of learning, trying, stumbling, and experimenting your way forward. Self-trust is the foundation that makes every other skill possible. Without that trust, women will sit outside the rooms where AI's rules are set, leaving the system to be built without their insights. It steadies your hand when the technology feels unfamiliar, and it keeps you curious when perfectionism tries to shut you down. AI literacy doesn't begin with knowing the right prompts. It begins with knowing you can figure them out.

The Old Narratives Holding Us Back

In my first book, *Glass Ceilings and Sticky Floors: Shatter Limiting Beliefs and Toxic Behaviors to Uncover Infinite Possibilities*, I introduce "SNAP," a four-step science-backed method that helps women break free from what I call "sticky floors." One of those four steps works to uncover the deeply held beliefs we have about ourselves, and it opens up the possibility to pivot out of them.

I want you to think about that little voice in your head. What's it saying? Is it saying, "I'm not techie" or "This is totally out of my wheelhouse"? Where did that voice come from? Often it's a product of the lifelong messages and experiences that we have gathered over time. Maybe you didn't see many women in tech roles as a little girl, or a teacher implied that boys were naturally better at computers. Over time, these hints collect into an internal narrative of not being "meant" for technology. And when we internalize these stories, we often remove ourselves from text design tables, which allows the default assumptions to endure.

Personally, I never identified as a "techie" person. I often joke that I married a technical project manager so I wouldn't have to figure out how to connect my modem and router if I moved. I never took a computer class outside of typing, and I tend to claim incompetence or ignorance if a tech question lands in my lap. I'm not alone. Research shows that women often underestimate their abilities compared to men with similar performance.[25]

In other words, many women feel that they have to know **everything** before stepping forward, while men assume they can figure it out as they go. But this isn't because we lack skill, it's because we have been conditioned to play it safe and avoid risks unless we are OVER-PREPARED. This keeps us small, but even worse, it results in holding ourselves back from learning opportunities, whether that's a new job or a new technology, unless we feel PERFECTLY ready.

Nowhere is this more evident than in the world of AI and emerging tech. It isn't the lack of talent or intelligence that keeps anyone out; it's the story in our heads. If you tell yourself you are behind or that you can't keep up with the changing trends, you are deciding to end the story before it even starts.

A large piece of my AI journey starts with my friend Greg Boone (mentioned earlier in the book), who asked me to be a co-host on a podcast called *AI Voice or Victim*. Greg is a CEO, investor, and an ally and advocate who is loud about supporting women, and he saw something in me that I didn't see at the time. When he approached me, I almost laughed in his face, because the first thought I had was "Who am I to talk about AI?" Nevertheless, he persisted, and he kept asking until I gave in. That decision put me inside conversations shaping AI, illustrating how saying "yes" places women in positions to influence the technology. *And who am I NOT to be talking about AI?* The truth is, I started using ChatGPT in

2020 when it first became "mainstream." I was experimenting and learning to create different GPT agents. But because it wasn't perfect, and I wasn't speaking the AI lingo, I didn't consider myself someone worthy of speaking on the topic of AI generally, let alone leading a weekly podcast. Once I explored my own narrative and examined my limiting beliefs around tech, I realized that I felt this way because I had no formal education around technology (which made me feel unprepared), and I had very few female mentors in the tech space.

The reality is women have always been a part of technology since its very inception. Ada Lovelace, for example, is cited as the world's first computer programmer back in the 1840s. Note, this isn't just the first *woman* computer programmer, but the *FIRST* computer programmer, and since 2009, she's been recognized annually on October 15th to highlight the often overlooked contributions of women to math and science.[26] Her legacy shows that women have always been technology's architects. We must claim that role again in AI's next chapter.

Women were literally among the first to write code and program machines. So, the next time you question if you belong, I want you to remind yourself that you do belong here. You don't need permission. We have always belonged in the tech space. The real challenge is convincing yourself of that fact, and silencing the inner critic that says you don't. History affirms our place in technology. Confidence, however, remains one of the most persistent barriers for women entering and advancing in technology and AI.

The Confidence Gap in Action

The research on confidence is never-ending. For decades, psychologists, neuroscientists, and leadership experts have debated whether confidence is something you are born with or something you build.

"Women were literally among the first to write code and program machines. So the next time you question if you belong, I want you to remind yourself that you do belong here. You don't need permission. We have always belonged in the tech space."

Not surprisingly, there was an entire book written on this very topic titled *The Confidence Code* by Katty Kay and Claire Shipman.[27] In this book, the authors set out to explore why men, despite women being better educated and more qualified in many cases, still dominate in leadership, pay, and promotions. Their findings showcase that confidence, not just competence, plays a central role, and this is a theme I explore in great detail in my first book.[28]

Studies on gender and tech consistently reveal a "confidence gap," with women often rating themselves lower and hesitating more, even when their skills are on par with men's. In my experience coaching women leaders, that hesitation doesn't come from lack of ability but often from the uncertainty around new or unfamiliar tools.

In the context of AI at work, this can mean women hesitate to try new tools without explicit permission or fear of looking foolish, leaving the experimenting (and the learning that comes with it) to men. At this point in time, multiple studies indicate a significant gender gap in the use and acceptance of AI technologies. In general, women have been slower to adopt AI tools than men, though recent trends do show this gap may be narrowing. For example, a Harvard-led working paper synthesizing data from 18 studies (140,000+ participants worldwide) found that women's adoption of generative AI was about 25% lower than men's on average.[29]

Surveys across diverse countries echo this pattern. A 2024 Federal Reserve Bank of New York poll reported roughly half of men had used generative AI in the past year, compared to about one-third of women.[30] Deloitte's global consumer research likewise noted that in 2023, only 11% of women in the US had experimented with generative AI versus 20% of men. By 2024, usage rose to 33% women versus 44% men, still a double-digit gap.[31]

While these numbers aren't where we would like them to be, encouragingly, Deloitte projects that women's experimentation with generative AI will accelerate and could reach parity with men by 2025 in the U.S., given the rapid uptick in women's usage in the last two years.[32] However, at the time of writing, men remain more frequent early adopters in many settings.

If you're curious as to why women appear to be leaning away from these technological advances when men seem to be leaning in, the data suggests that women are being more thoughtful about it, or in other words, they don't trust it yet. While men tend to dive in with curiosity and confidence (or a fake-it-till-you-make-it attitude), women often approach AI with more caution, asking the bigger questions:

- Is this ethical?
- Is my data safe?
- Will I be judged for using this at work?

These questions are crucial, but if caution keeps women from engaging at all, the resulting systems will reflect only those who dove in, believing the reward is worth the risk.

A large international Pew Research Center survey conducted in 2025 across 25 countries found that men are "more likely than women to have heard a lot about AI" and that "women are more likely than men to be mainly concerned about the increasing use of AI."[33]

While all of those questions and concerns are valid, and should be reviewed carefully and with intention, that hesitation adds a different kind of risk. That risk could add up to missing skill-building on key technologies, visibility, and impact.

One peer-reviewed study in *Frontiers in Psychology* (2025) tested gender differences in AI-related perceptions and found that women report higher AI anxiety and lower positive attitudes toward AI compared to men.[34] Women in this study also self-reported lower usage of AI and lower perceived knowledge of how AI works.[35] These findings suggest a confidence and comfort gap that I will argue directly hinders women's engagement with AI tools.

One key factor identified is a "trust and ethics" gap. Women users tend to have less trust in AI systems and more ethical reservations. Deloitte observed that despite women's accelerating uptick of generative AI, they express significantly less trust than men that AI providers will keep their data secure.[36] This trust deficit can reduce women's willingness to use AI regularly.

Similarly, Harvard Business School Associate Professor Rembrand Koning conducted research that women often question whether using AI tools is ethical and fear being judged for relying on AI in the workplace.[37] These concerns can make women hesitate to adopt AI even when it offers incredible, never-seen-before productivity benefits. The *Harvard Business Review* working paper warns that if women continue to turn away from AI due to these worries, they risk missing out on skill development and productivity gains, potentially widening gender gaps in career advancement. That widening gap isn't just individual; it shifts who controls how AI is used in hiring, promotions, and pay—the structural levers of power this book is about.

What's noteworthy is that two things can influence these trends: **exposure and encouragement**. In contexts where women have strong tech backgrounds or high levels of support, the gap can shrink, or even reverse. Take, for example, the support I received from Greg Boone, who encouraged me to lean into AI even further and be his co-host on our podcast. Greg signaled that my voice belonged in this space (even if I wasn't sure of it myself). That direct encouragement is what gave me the courage to say yes and step fully into AI.

The same pattern shows up in the data. In one survey of tech workers in San Francisco, women were slightly more likely (3%) than men to use AI.[38] This population, focused in the California Bay area, had extraordinary exposure to AI tools, early-access technologies, on-the-job training, daily use, and worked in environments that **actively encouraged experimentation rather than punishing imperfection.** Because access and encouragement were built into the culture, the usual adoption barriers looked very different, and when women are exposed and encouraged, they not only adopt AI at higher rates but also co-design its features, proving that disparities are created by context, not capability.

When the baseline conditions are leveled in this way, women's adoption rates not only rise **but they can surpass men's**, suggesting that adoption disparities are not inevitable, but that they are shaped by environment. (You should know that these results are an outlier compared to the broader research, but the concentration of tech talent, resources, and experimentation culture in the Bay Area helps explain why exposure and comfort with these tools are markedly higher.)

Overall, however, the consensus of global data at this moment in time is that women on average use AI less frequently and less confidently than men, largely due to differences in trust, awareness, and perceived relevance. If we address these issues of confidence and trust, we begin closing the gender gap in AI adoption.[39] This matters because early adopters shape the rules, norms, and ethical boundaries of new systems; if women aren't among them, their perspectives won't be built into the system.

Early adopters gain skill velocity. Skill velocity increases visibility. Visibility drives promotion. Promotion drives pay. The adoption gap is not about comfort with technology. It's about compounding advantage over time. In other words, those who adopt first get to write the playbook. If women aren't in that cohort, AI will be optimized for someone else's experience.

"Early adopters gain skill velocity. Skill velocity increases visibility. Visibility drives promotion. Promotion drives pay."

Self-Trust: The First Form of AI Literacy

It turns out the key to closing these gaps isn't simply learning more tech skills, getting an AI Certification, or joining the AI Committee at work (although those do help!). The key to closing these gaps is learning to trust your ability to learn.

Each December, I work on picking a "Word of the Year," which has become a popular practice when planning your upcoming year's goals. Every year, I struggle to land on the perfect word, and I often take several months into the new year to figure it out. Previous words have been COURAGE, GROWTH, and INTENTION, but for 2025, I selected the word TRUST.

At the time, the meaning was deep and personal, and related to trusting my instincts. This would be my first full calendar year in business, and I wanted to trust my gut on more of my decisions. But what I've learned over the years is that it's so much more than just trusting my gut. It's trusting my knowledge, trusting my skills, trusting the universe that I'm going in the right direction (If you couldn't tell, I'm a little woo-woo). But what I've learned is that you must treat your self-trust as your internal tech primer. Before you can leverage any AI tool, you have to believe that you can truly figure this out, and this statement is backed by research.

A recent study on AI adoption found that self-efficacy in AI learning, a.k.a. confidence in your ability to master AI, dramatically reduces stress and burnout when using technologies.[40] In the study, employees with higher self-efficacy experienced much less job stress related to AI implementation than those who doubted their abilities. Simply put, if you believe you can handle the new tool, you're far less likely to feel overwhelmed by it. Confidence literally acts as a shield against frustration and anxiety.

This is nothing new. When we trust ourselves to learn, we approach challenges in a much different manner. We approach them with possibility. In my coaching practice, I have found that women who don't have that trust often interpret any setback as proof that they aren't good enough. Their limiting beliefs bubble to the surface and take over, leaving them paralyzed with anxiety.

As you move through this book, I want you to think of self-trust as your foundation. I want you to believe in your ability to take on any AI tool or technology that interests you, and I want you to truly believe you can master those tools. Because no amount of certifications or classes will stick if deep down you have already decided you're going to fail. Henry Ford said it perfectly, "Whether you think you can, or you think you can't, you're right." Believing you can learn AI isn't just about personal growth. It's about stepping into positions where you can ensure the technology serves everyone.

The good news is that self-trust is like a muscle you can train and build over time. Self-trust is a mindset you can practice and work at each day. In my book, *Glass Ceilings and Sticky Floors*, when we are stuck in the sticky floor of self-doubt, we pivot. "**YET**" is one of my favorite words. "I haven't mastered this, *YET*." That "yet" implies that in the future you will have done this. Once you say it with conviction, you are on your way. That pivot is powerful. It transforms AI from an anxiety-inducing stressor into an opportunity.

Rewriting Your Narrative: From Fear to Fun

So how do we actually change the story that we tell ourselves about technology? That was the question I had to dig down deep to ask myself, because the belief that I wasn't "techie enough" was a story I'd told myself for years.

It starts with small, intentional shifts in perspective. In the pages that follow, we'll walk through the transformations, and I hope this book inspires you to move from anxiety to curiosity, hesitation to experimentation, and fear to fun. Consider this your personal coaching session with practical steps that will help you see what's possible.

From Anxiety to Experimentation

For many women, the biggest barrier to using AI is the anxiety that shows up when they start thinking about learning and implementing AI in a way that's meaningful. That tightening in your chest, the fear of "doing it wrong," is the same fear that fuels hesitation. But if we're to push through this crushing anxiety, it will be because we're harnessing the power of a curious mindset. Curiosity softens the pressure to be perfect and not make any mistakes. And once you let yourself be curious by clicking around and testing ideas, you naturally step into experimentation. With each experiment you conduct, you build familiarity, confidence, and momentum.

AI THAT IS PERSONAL, SAFE, AND FUN!

When Greg Boone, my podcast co-host and business partner, and I teach AI, we always do it through this simple lens: make it personal, make it safe, and make it fun. When organizations adopt these three principles, the intimidation around AI drops, people start experimenting, and, dare I say, laughing and having a blast playing with AI. I want you to make it personal, make it safe, and make it fun for you too, and here is why:

PERSONAL: AI just "hits different" when it's connected to something that you actually care about. When you give it a problem, a stressor, or a goal from your real life, you start to see how it can solve these problems in real time. You feel the relevance it has, and any resistance you might have been harboring starts to vanish. I call this the "ah-ha" moment—that flash of clarity when something finally clicks into place.

SAFE: I always want you to start with baby steps. You have to walk before you can sprint, so no feeding the AI tool all your bank transactions from the past 12 months. However, you could ask it to act as your personal money coach to help you develop an abundance mindset (something I have done!) and to help you rewrite the stories you tell yourself about money. Keep the stakes low until you have a deeper understanding of how the tool works.

FUN: Fun keeps the pressure off and your perfectionism away. One of the first things I ever did with AI was teach it to talk to me as if Matthew McConaughey met Gerard Butler (IYKYK)! Play invites experimentation, and it reminds your nervous system that this is safe. And as Matthew McConaughey would say, "Alright, alright, alright!"

From Fear to Fun

The strongest leaders turn fear into play. By doing so, they stay in the arena where AI's norms are set, instead of stepping back and letting others define them. Yes, AI is serious business, and there's a lot to learn, but who says we can't make it a fun and interactive process? Fear comes from imagining worst-case scenarios and believing those will come true. You may think, "If I mess up with this AI, I'll ruin the project, or look stupid, and even risk my reputation!" Yikes! And while it's true that we should be thoughtful with a tool as powerful as AI, we often inflate the stakes in our minds, and the antidote to reducing the fear is to introduce a sense of play.

Play is powerful because it lowers our inhibitions. While you are having fun, you aren't as self-conscious or afraid of looking foolish. One example of "making it fun" is when I experimented with ChatGPT to create coloring pages of my daughter's seven-year-old cheerleading team for the girls to color. It took several different prompts to get it just right, but the outcome was fun for ALL! When you turn your fear of AI into fun *with* AI, you drop the heavy weight of "this is scary," and you replace it with lightness and even a little laughter!

Embrace Your New Narrative

Rewriting your story in this way is liberating. It's like upgrading your own internal operating system. You're no longer running on the old program of doubt and intimidation. Instead, you are operating on curiosity, courage, and yes, even FUN. Remember that mindset determines how boldly you'll do anything, and when you believe that technology is *for you*, and that you're capable of learning it, using it, and using it well, a world of opportunities opens up. Those opportunities aren't just about using AI; they are about influencing how AI is built and deployed, which is where power resides.

The next time you encounter a new AI tool, or you hear the phrase "vibe coding" and think, "Oh my word, what is that?" pay attention to your inner dialogue. If the old whispers of "I'm not cut out for this" surface, I want you to consciously choose a new narrative. One that says, "I might not know this yet, but I can learn." Because you can, and you will.

In the journey ahead, every chapter will equip you with knowledge and practical techniques to thrive in an AI-powered world. But none of it will matter if you don't first give yourself the permission to dive in headfirst. I want you to own your narrative. I want you to change how you see yourself in relation to technology, and I want you to challenge the story you have told yourself that you aren't "techie enough" or "smart enough" or "young enough" to learn this.

AI will integrate into hiring systems, promotion pipelines, compensation models, and productivity expectations regardless of individual readiness. These shifts are already underway. Your internal narrative sits inside that infrastructure, influencing how you engage with visibility, authority, and opportunity as these systems evolve. You are not an outsider to this transformation. You are already a participant in it. When you step in as a rightful architect of AI, you help redesign the system itself, preventing old inequities from being hard-coded into the new order. The systems will scale. The only variable is how you position yourself within them.

TLDR

- AI will scale the systems that already exist. If those systems remain unexamined, their defaults will be amplified at speed.

- Your internal narrative functions as infrastructure. Self-trust influences experimentation, exposure, and visibility. Those factors determine whether women are among the architects of AI or merely its users.

- The gender adoption gap is not only about skills. It reflects structural conditions around trust, encouragement, and access.

- How you position yourself within AI systems matters more over time. Engaging early increases your exposure, strengthens your credibility, and opens new doors.

YOUR AI-POWERED ACTIONS

ASK:

What story have I been telling myself about AI or technology?

APPLY:

Apply the Pivot. This is the "YET" rewrite. Write down one limiting thought you have had about tech, AI, learning, or leadership. And add the word "YET" to the end of it. Then underneath, write what "it is done" looks like (the future version of success). Example:

- "I'm not confident in my AI capabilities... YET."
- "It is done. I use AI every day with confidence to make my work easier, and my impact stronger."

AMPLIFY:

Build momentum! Share one small win from your AI exploration with a peer or your team. Remember, visibility builds confidence (and is incredible for your career too!).

CHAPTER 2

VISION ENGINEERING WITH AI

When I was building HER Collective, I had no roadmap, no checklist, and no idea what I was doing. Today, it's a leadership community where ambitious women gather to accelerate their careers, strategic power, and influence in the corporate world.

But at the beginning, I couldn't quite see the full shape of it, and every time I sat down with my favorite notebook and pens to map it all out, I would just stare at the blank page. One morning, partly out of frustration and partly because I needed some real momentum, I opened ChatGPT and started asking it to help me map out the idea in my head. Immediately, as the words started appearing on the page, the fog lifted, and I was able to see my vision taking place, right before my eyes—not just as I'd described it, but better. It was clarified, expanded, and elevated beyond what I could have imagined.

That was the moment I started getting genuinely excited about HER Collective, and the potential it had to change the lives of so many women, but it was also the moment I realized that AI had a place in my world going forward. If it could help me engineer my vision in minutes, it could help the women in HER Collective do the same, without forcing insight onto a blank page or waiting for confidence to arrive first.

AI turns inspiration into design at scale. When you introduce AI into a workflow, you're not adding a tool; you're installing infrastructure. What I learned in that moment wasn't just how powerful AI could be. What struck me was how much more powerful human intuition becomes when it's supported. I began to see AI as something that sharpens how I build, lead, and imagine what's possible. Infrastructure scales patterns. Whatever logic sits underneath your decisions, clear or unexamined, is what AI will scale. This chapter explores that idea.

AI, even at its very best, does not replace human vision. It refines it. It sharpens what you already sense, amplifies what you already value, and helps you see your intuition more clearly. Just as a strong leader uses systems to scale culture without losing its heart, AI becomes a mirror to human intention, not a substitute for it.

What Vision Engineering Really Is

Vision engineering is the intentional process of designing your future before it happens. It's about creating clarity around what you truly want, being intentional, and then using tools, both human and technological, to help you get there. It's the art and science of building both a mental and a digital model of your desired future.

Vision engineering with AI is all about harnessing the tech as a tool to **reflect and refine your goals**, rather than expecting it to magically predict your future. At its core, it combines **intuition** (what you feel drawn to), **insight** (what your experience and data reveal), and **intelligence** (both human and artificial) to help you visualize, refine, and reverse-engineer your goals into reality.

AI doesn't decide the vision. You do. And whatever you decide, intentionally or unconsciously, is what it will scale. That truth becomes powerful when we look at what happens inside real organizations.

Several years ago, Amazon built an AI-powered hiring tool to streamline its recruiting process. The goal sounds practical, efficient, and data-driven. The system was trained on 10 years of historical hiring data to identify the patterns of candidates who had been successful in the company. But history has a memory, and the data it learned from reflected a workforce that had been overwhelmingly male. The algorithm began penalizing resumes that included the word "women's," as in "women's employee resource group (ERG)" or "Women's Engineering Society," and it downgraded applicants from women's colleges. It learned unintentionally but precisely that male-coded experience was the norm.[41] Anything else was an outlier.

The AI wasn't intentionally malicious; it was obedient. It mirrored the vision embedded in the system that trained it. Amazon ultimately scrapped the tool, but the lesson is not about one company's mistake. It's about something far more important.

AI doesn't introduce bias from nowhere; it amplifies what's already there. That's why vision engineering matters. Because if you don't deliberately embed equity, safety, and inclusion into the system at the design stage, AI will faithfully reproduce the narrow patterns it was trained on. The machine is obedient. The responsibility is ours.

This isn't a new phenomenon. Every major technological shift has reflected the priorities of the era that built it. Industrial systems amplified industrial hierarchies. Financial systems scaled the assumptions embedded in their institutions. Digital platforms accelerated whatever incentives and blind spots were coded into their foundations. As much as we would like for AI to break those patterns, it's simply accelerating them.

In the nineteenth century, every American town kept its own local solar time. Noon meant when the sun reached its highest point in that specific place. And that worked until railroads connected cities across long distances. Suddenly, trains required synchronized schedules. Local time became operational chaos. In 1883, major railroad companies imposed standardized time zones across the country before the federal government formally adopted them in 1918[42]. Infrastructure reorganized reality. Time itself became systematized.

"AI doesn't decide the vision. You do. And whatever you decide, intentionally or unconsciously, is what it will scale."

AI operates the same way. Once embedded into decision systems, it standardizes judgment. It turns local discretion into scalable logic. The question isn't whether it will coordinate behavior. It will. The question is whose logic becomes the standard.

That answer isn't decided at a later time, but in the design. AI will take whatever is already present and make it consistent, repeatable, and scalable. And it won't question it. Which means inputs matter. The assumptions matter. The vision underneath it all matters. Because once a system is built, it becomes a process, and it becomes part of how decisions get made, long after you've stepped away.

That's the responsibility. And the opportunity. You aren't waiting on the future to take shape, but you are actively defining the logic it will run on. What you build into it now is what it will scale.

The Mirror vs. The Crystal Ball

When people first hear about using AI for personal visioning, they often imagine a crystal ball, this all-knowing mythical machine that will spit out predictions or answers about their future. But that isn't the reality (and honestly, where is the fun in that?).

Instead, I want you to think of AI as a mirror. And just like a mirror reflects your appearance back to you, so you can adjust a stray hair or the lipstick on your teeth, AI can reflect your ideas, goals, and plans in a rich visual or structured way so you can see them more clearly.

A mirror reflects whatever stands in front of it. If you walk up to it rushed and distracted, that's the version you will see. If you approach it with intention, clarity, and purpose, that's what it gives back. AI works the same way. It doesn't invent your vision; it reflects the inputs, assumptions, and priorities you bring to it. Think of it as your partner in imagination and planning instead of a fortune teller.

This distinction matters. Your vision for your life or business isn't something that an algorithm can hand to you wholesale. It's something that's deeply human, deeply personal, and rooted in your values, desires, and your intuition. What AI can do is help you engineer that vision, render it in vivid detail, explore different scenarios, and identify blind spots or opportunities you might miss on your own. AI makes your internal vision visible and concrete, but you're still the architect of what you build. And whatever blueprint you formalize through it becomes repeatable, and once formalized, it becomes process.

Vision Engineering in Action

Before we go deep into the tools or techniques, I want to ground us in reality. Vision engineering shows up wherever a human has a clear concept or idea for what should exist (they feel it in their bones!), but they need help translating it into something others can see, understand, and act on.

From Mockup to Masterpiece

Meet Maicie Stern, a woodworking artist and small business owner with a passion for designing custom wooden accent walls. I met Maicie through my daughter's cheer team, and what struck me immediately was not that she used AI, but how naturally it fit into the way she already thought and created.

When we think of AI, we think tech, desk jobs, and click-clacking on the computer, but the reality is, there's a place for AI in nearly anything. Maicie is the founder and owner of The Happy Wall Company. She has an uncanny intuition and talent for design. She can walk into a room and FEEL what kind of statement wall would bring it to life. But as any designer knows, not every client can see the vision that's in your head. This is where AI entered Maicie's process, as a tool to help her translate that uncanny intuition into a vision her clients could see. What changed was not her creativity, but her workflow. AI transformed a subjective pitch into a structured, visual decision system that could be replicated with every client.

One day, sitting together at a cheer practice, I wondered how Maicie might use AI in HER work. We chatted about the possibilities. I suggested she download ChatGPT and use it to help her with measurements and systemizing quotes or contracts.

Maicie took it a step further.

Maicie had a bold idea for a current client—a sweeping geometric pattern that would be a bold statement wall. But her clients were on the fence. They liked her portfolio, but they weren't sure how the concept that was living in her head would look in their space, in their HOME. Normally, Maicie might sketch her design by hand or create a mood board and hope the client "got it," but this time, she tried something new.

Maicie turned to AI. She fed the AI a photo of the clients' room, specifically the wall, she fed it measurements, and the client preferences (their taste, current home vibe, etc.), and then she described her vision in detail. "Show a floor-to-ceiling wooden geometric accent wall..." In less than two minutes, AI rendered an image that was astonishingly close to what Maicie had imagined. It was as if the idea in her mind had leapt out onto the screen in front of her. In that moment, her design process shifted from persuasion to prototype. The decision was no longer based on imagination alone, but on a shared model.

When Maicie presented these AI-generated mockups to her client, their eyes lit up. The vague uncertainty was gone, replaced by concrete excitement: "This is going to be our accent wall!" In that meeting, Maicie experienced firsthand that AI was acting as a mirror for her creativity. It took her intuition and artistic idea, and reflected them back to her in a visual form that others could see and understand. AI didn't

"This is where AI entered Maicie's process, as a tool to help her translate that uncanny intuition into a vision her clients could see. What changed was not her creativity, but her workflow."

come up with the design—Maicie's heart did that—but AI amplified her vision by making it visible and adaptable in real time.

In the end, the AI-generated mockup helped Maicie land the project. What she'd built was not just a rendering. She'd created a repeatable architecture for client alignment. In fact, the clients said that seeing the concept in their actual home versus a typical mockup was the deciding factor. It gave them confidence—a kind of future memory of how the finished wall would look and feel inside their home.

Maicie went on to build the full installation, which surpassed even the original mockups, and the clients were thrilled with the result. Not only that, but Maicie's business has grown too, and her referral business is booming, with previous clients often mentioning how high-tech and smooth her design process is.

For Maicie, the experience was a revelation. She later shared, "I always trusted my gut and my eye, but sometimes it was hard to persuade clients to trust my vision, or to pick a vision." Now, Maicie can show them what that means, and her clients can see it with their eyes instead of having to imagine it in their minds.

These AI mockups haven't replaced her hand-drawn sketches or her material experiments, but they do sharpen her presentation to clients. She still gets to decide what to create, but now she has an amazing assistant that can powerfully illustrate her intuition. This is vision engineering in action—using AI to bridge the gap between the idea in your mind and the reality you're working to create. And when that bridge becomes part of your operating process, it stops being a creative assist. It becomes infrastructure. Every future client interaction now runs through that system. The result is not only a clearer path forward but also a stronger conversion in following it (and in Maicie's case: more clients and more revenue).

Why Visualization Works (and Why AI Helps)

Maicie's story highlights something else, too: the emotional boost that comes from visualizing your goals. When she saw her idea rendered realistically, it wasn't only the clients who were convinced. Maicie herself felt a surge of confidence. *Yes, this will look incredible. I can't wait to build it.* That motivation is priceless. It's the kind of intrinsic excitement that fuels follow-through. And science backs this up.

Research in psychology has found that vividly imagining a successful future can increase motivation and persistence because you begin to experience the joy and accomplishment of that future in advance.[43] In Maicie's case, seeing the AI mockup was like peering into the future. It made the outcome tangible, and it made it inevitable, which in turn made her even more committed to making it happen.

You might ask, how can visualization do all this? You might feel this is all fluff, but the truth is, visualization helps because it creates:

- **Clarity** around your goals, making them specific and actionable.
- **Motivation** to get going! You start to feel the reward before you even begin!
- **Future-self Connection.** This one is important because you care more about the person you are becoming than the one you are today. Establishing that connection is key to momentum.
- **Problem-solving Readiness:** You mentally rehearse any obstacles or challenges before they show up!

AI enhances this process by making visualization tangible. Instead of holding everything in your head (or one of the thousands of notebooks you bought yourself with the dream of being more organized), you can externalize it through images, scenarios, or written drafts and engage in real time.

See it. Script it. Then Prompt It

Vision engineering works best when it follows a clear order. When we skip steps or blur them together, we create confusion for ourselves and the AI tools meant to support us. Here's the sequence I find to be the most effective.

First, you **see it**. Before you open a notebook or type a single line, close your eyes and allow yourself to imagine the future clearly and honestly. (Now, do it again, but let's dream bigger, OK? No holding back!) Vision always starts with you. Picture what success looks like (not just the outcome, but the feeling of being there!) Where are you? What are you doing? Whom are you serving? How does it feel in your body? This step is intuitive, emotional, and deeply human. There's no AI involved yet. This is about letting yourself *want what you want* without editing, shrinking, or explaining it away.

Once you've imagined it, then you **script it**. You need to write it down. Scripting your vision means translating that mental picture into words, on paper or on a screen. Writing your vision down transforms it from an inspiring idea into an ironclad commitment. Research shows that if you write down your goals and you revisit them on a regular basis, you are significantly more likely to follow through on those goals.[44] The act of scripting forces clarity, creates accountability, and bridges the gap between present you and future you. At this stage, you are still leading. You are naming what matters, and you are deciding on your "why."

Only then do you **prompt it**. Only after you visualize and script your vision do you bring AI into the process. This is where AI becomes truly powerful as a refiner and accelerator of your vision. With a written vision in hand, your prompts can become sharper, and you move from vague goals to precision coaching. Instead of asking AI, "What do I do with my life?" you can ask:

- Help me pressure test this plan.
- Show me three paths to reach this goal.
- Help me structure this vision into a 12-month roadmap.
- How can this vision integrate with the real rhythms and responsibilities of my life?

AI doesn't replace your thinking here; it amplifies it. Your intuition sets the direction. Your written vision provides the foundation. AI adds structure, rigor, and speed.

If you follow this order—see it, script it, then prompt it—you stay in the driver's seat while letting AI do what it does best: expand, clarify, and accelerate what you already know to be true.

The Power of Prompts: Shaping the Vision

It's worth emphasizing that in this dance with AI, the prompts you give it are incredibly important. In the prior chapter, we said that self-trust is your first form of AI literacy. It's my opinion that the second form of AI literacy that must be mastered is the power of the prompts, because one of the unexpected benefits of engaging with AI is that it forces you to frame your thoughts.

When you sit down and ask AI a question or ask it to assist with a task, such as "Help me map out a six-month plan to launch my podcast, given I work full-time, have two kids, and can only dedicate 10 hours a week," you are essentially coaching yourself into clarity. But there's a method to the madness. When my business partner, Greg Boone, and I teach the basics of AI, we focus on the Five Steps to a Perfect Prompt:

1. **Role:** Who do you want AI to be?
2. **Goal:** What do you want AI to achieve?

3. **Context:** What's your reality?
4. **Constraints:** What boundaries matter?
5. **Confirmation:** What do you want it to clarify before it starts?

These steps are simple, powerful, and endlessly adaptable, whether you are building a brand, planning a thought-leadership article, or mapping your next career leap. This is how you move from the random back-and-forth within AI to precision coaching, and that will get you the clarity you need to get started.

Role: Who do you want AI to be?

You would not ask a lawyer to design your logo or a designer to negotiate your contract. The same thing goes for AI. Defining the role forces the system to adopt a specific lens, skill set, and decision-making framework instead of offering generic advice. When you assign a role, you elevate the output from surface-level response to expert-caliber thinking.

For example, "Act as a world-class executive coach that specializes in women who are launching their first podcast."

Goal: What do you want AI to achieve?

AI cannot read ambition. It can only follow directions. A clearly defined goal anchors the conversation in outcomes, not ideas, and prevents the spiral of vague productivity. When you state the result you want, you shift from exploration to execution.

For example, "Help me build a six-month plan to launch my podcast."

Context: What's your reality?

Context is the difference between advice that sounds smart and advice that you can apply and actually works. The more AI understands your season of life, resources, limits, and experience level, the more it can tailor the strategy to you instead of an imaginary ideal user. Context humanizes the machine.

For example, "I work full-time, have two kids, and can dedicate 10 hours a week."

Constraint: What boundaries matter?

Constraints create quality. They narrow the field of options and force precision. By defining what must be included, excluded, simplified, or limited, you turn AI from an idea generator into a strategic partner operating inside real-world limits.

For example, "Keep the plan simple and family-friendly. Avoid jargon. Limit each phase to four to six tasks with estimated weekly time."

Confirmation: What do you want it to clarify before it starts?

This is my favorite, and it's often the secret sauce. Instructing AI to pause and ask clarifying questions ensures alignment before execution, preventing wasted effort and surface-level answers. Confirmation transforms the interaction from command-and-response to co-creation.

For example, "Before you begin, ask me up to three questions, one at a time, to clarify my goals and timeline."

That single sentence sets two boundaries:

1. It sets the limit so that it doesn't spiral into endless Q&A;
2. It provides the pacing because AI learns from your answers as it goes, and giving it responses one at a time helps it refine in real time.

The reason this is so important is that, as humans, we often have more context in our brains than we can actually describe. This is what Polanyi's Paradox points to: we know more than we can say. This step gives AI the opportunity to ask those questions and clarify what it is you really want.

FIVE STEPS TO A PERFECT PROMPT

When you bring together the five steps of a perfect prompt, you turn AI from a guessing machine into a thinking partner. Instead of hoping for the right answer, you are teaching (and guiding) it how to work with you.

ROLE: Act as a world-class executive coach who specializes in women who are launching their first podcast.

GOAL: My goal is to build a six-month plan to launch my show, *Glass Ceilings and Sticky Floors*.

CONTEXT: I work full-time, have two kids, and can only dedicate 10 hours per week. I have no idea what I'm doing or how to start a podcast.

CONSTRAINTS: Keep the plan as simple as possible. Each month should have four to six core tasks that fit within that 10-hour weekly limit.

CONFIRMATION: Before you create the plan, ask me **up to five questions, one at a time,** about my show, the concept, or anything else to help you gain clarity.

Before we go further, a quick note: the examples in this section are intentionally simple to teach the structure. My real prompts are often far more detailed. What matters most is that you understand the architecture first. Depth comes naturally once the foundation is strong.

At the end of the day, the prompts you feed AI shape your perspectives. I want you to think of prompts as your chisel, and your vision as the big block of marble that has yet to become the sculpture. When you start asking different questions, you carve the vision from new angles.

Do not be afraid to experiment with prompt engineering. Prompt Engineering is the technical term for how you talk with the machine. This is where the fun is! If one approach gives you something flat or forgettable, change the way you ask. Most people default to functional prompts like, "List the steps do to X." That will get you an answer, but not always one you feel connected to or motivated by. Try shifting the frame instead. Ask: "Tell the story of someone achieving X and show me how they did it."

Now you're not just getting the steps, but you're getting movement, emotion, and context. You might see where she struggled, what almost made her quit, and what pulled her forward. Anyway, you might read something like: "She almost gave up in month three, but what kept her going was remembering why she started." That kind of detail changes how you engage with the plan; it moves it from something you read to something you can actually see yourself inside of. The difference is subtle, but it matters. One version gives you a set of instructions; the other version gives you traction. This is the real power of prompt engineering. It's not just about getting an answer; it's about shaping the kind of answer that actually moves you.

Remember, the possibilities are endless.

Vision Engineering at Scale: Bumble

For those of you who have been in a committed relationship since 2014, you might not know what Bumble is (I didn't). Bumble is a social and dating app where women initiate the first move in a match. When Whitney Wolfe Herd founded Bumble in 2014 after leaving Tinder, she had a crystal-clear vision to create a dating app that was built on empathy, kindness, and women's empowerment. Whitney's guiding belief was that technology should serve human needs, and in this case, the need for women to feel safe and respected when dating online—not the other way around, as it historically is. This vision of a more human-centered tech experience became Bumble's North Star, and interestingly, AI played a key role as Bumble grew, not by taking over decisions but by scaling Whitney's vision and values across millions of users.

One powerful example is how Bumble tackled the all-too-common problem of unwanted explicit images—those infamous dick pics that unfortunately plague too many women online. Rather than shrugging it off as an inevitable nuisance, Whitney's team treated it as a core safety issue that needed solving to fulfill Bumble's vision of a respectful platform. They engineered a solution that combined user empathy with AI's pattern recognition power.

In 2019, Bumble introduced an AI tool called Private Detector, an algorithm trained to automatically detect and blur lewd photos before a user ever sees them. Astonishingly, the machine learning system could determine with 98% accuracy if an incoming image was pornographic or inappropriate.[45] If someone sent a nude or lewd image, Bumble would intercept it, cover it with a warning blur, and give the recipient the choice to view it or not, saving millions of women from dick pick delivery.

This is what AI looks like when it is guided by a very human value: decency. Whitney Wolfe Herd publicly stated that from the very beginning, her vision was to make the internet "friendlier to women." She partnered with her team to "use the power of technology and scale to do good things, keep people safe, and make us more accountable" rather than exploiting or harming users.[46] This is a perfect embodiment of AI as a mirror to human intention. The AI was "taught" what respect looks like, and what violations look like, according to Bumble's principles, and then it reflected that learning at scale to protect its users. Bumble's technology doesn't replace the need for users to treat each other with decency; it reinforces it by creating an environment where bad behavior gets flagged and good behavior is the expected norm.

The impact on Bumble's growth was significant. By making safety and respect a priority, Bumble built enormous trust, especially with female users who had been burned by other platforms. That trust translated into more users and into a distinct brand identity. Whitney's intuitive understanding of her users' needs, born of empathy and her own experiences, was amplified by AI tools that could enforce and scale the solutions to meet those needs. As a result, Bumble became one of the fastest-growing social networking companies in the world, and Whitney Wolfe Herd is the youngest female CEO to take a company public.[47]

But Whitney's use of vision engineering at Bumble doesn't stop at policing bad behavior. She's also spoken about how AI might enhance the core dating experience, not by taking the humans out of it but by helping people put their best selves forward.

In a 2025 interview, Whitney observed that many singles struggle with confidence in dating, saying the average US single doesn't date because they don't know how to flirt or they're scared they don't know how. She mused about leveraging AI chatbots as a kind of coach or practice partner, an AI that helps users craft messages or icebreakers, instills confidence, and gets feedback in a safe way.[48] Even as she explores these AI-driven features, Whitney Wolfe Herd is adamant that AI will never replace the human element of dating. She quipped in that same interview that if anyone truly thinks chatbots will replace the need for real human love, "We'll have bigger issues than Bumble."[49]

Bumble's philosophy illustrates a key theme of this chapter: AI refines our human intuition. It doesn't replace it. Whitney Wolfe Herd had a vision of empathetic online interaction, and she uses AI as a tool to mirror that vision back to users through safer features and smarter matching. But the empathy at the heart will always be human.

Both Big and Small: Engineering a Vision is for ALL

If Maicie shows us what vision engineering looks like in one room with one client, Whitney Wolfe Herd shows us what it looks like when those same principles are built into a platform used by millions. Through these two unique stories, we see vision engineering on both a small and large scale. Leaders define a vision and then use AI as one of the tools to engineer that vision into reality. It's a dance between future thinking and real-time adaptation that requires you to envision, implement, learn, and refine. Both Bumble and Maicie continue to use AI and monitor it within the business, and by doing so, they stay aligned with their guiding vision while navigating the complex, ever-changing terrain of technology and human behavior.

Maicie used AI to amplify creativity. Whitney Wolfe Herd used AI to scale empathy. Amazon unintentionally used AI to replicate an old baseline. Same technology. Different vision.

The best part is you can use these same principles in your own life and work. You don't need to be a tech CEO to apply vision engineering. Whether you're planning a career move, writing a book, or trying to improve your health, the combination of clear vision, empathy (for yourself and others), and AI assistance can be transformative.

"AI doesn't wait for you to have perfect clarity. It embeds whatever clarity or confusion already exists. When you introduce AI into hiring, pricing, evaluation, creative production, or strategy, you're installing a decision architecture that will operate at scale."

AI doesn't wait for you to have perfect clarity. It embeds whatever clarity or confusion already exists. When you introduce AI into hiring, pricing, evaluation, creative production, or strategy, you're installing a decision architecture that will operate at scale. If the architecture is vague, it will scale vagueness. If it's biased, it will scale bias. If it's intentional, it will scale intention. Vision engineering is not about motivation. It's about being intentional and designing the system that lives behind your decisions.

Engineering Your Vision, One Step at a Time

By now, you have a toolkit of techniques, both human and high-tech, to bring your goals into focus and move boldly toward them.

I want you to imagine yourself right now as the vision engineer of your own life. You have blueprints (your visualized goals), you have tools (AI assistants, prompt strategies, and good ole pen and paper), and you have that inner fire (your intuition, desire, and ambition). I want you to take a moment and revel in that. It's empowering, isn't it?

The future isn't some foggy uncertainty that you passively wait to see what happens; it's something that you actively design and build. After all, you are the architect of your own life.

As you finish this chapter, I want you to consider one goal or idea that has been lodged in your heart and on your mind. How could you use what we've discussed to accelerate it? Maybe tonight you'll sketch out a vision board or ask an AI to mockup that business idea you've been toying with. Maybe you've finally written down a personal mission statement and your core values. Whatever it is, do one small act of vision engineering for yourself. Even if it's simply closing your eyes and picturing that future in rich detail for five minutes. That's a start. You are literally beginning to engineer the neural pathways for success in your brain, and once operationalized, they can be embedded into systems.[50]

Maicie showed us that AI can make vision visible. Whitney Wolfe Herd showed us that AI can make values scalable. In both cases, the technology didn't create the outcome. It extended what was already there. Your vision is the irreplaceable star of the show and the starting point. AI is the system that carries it forward.

Now it's time to take this energy and insight and carry it forward. In the next chapter, we move from reflection to architecture. We have to get our hands dirty! You'll learn how to use AI not just to imagine what's next, but to design the systems that will shape it. You'll see how AI can help you map your next promotion, position your value, identify growth opportunities, and build a strategy that compounds over time.

Because the moment you move from imagining to building, your vision stops being personal. It becomes structural. It begins shaping teams, decisions, promotions, and policies. And once AI operationalizes those patterns, they harden in place, becoming impossible to change.

Infrastructure scales. You've built the vision. Now we engineer the path.

TLDR

- Vision engineering is designing your future on purpose and then building toward it with clarity.

- AI is a mirror, not a crystal ball. It reflects and refines your goals. You remain the architect.

- Human judgment stays central. Use AI to add rigor, speed, and structure.

- Prompts drive quality. Use the Five Steps to a Perfect Prompt (Role, Goal, Context, Constraints, and Confirmation) to get tailored, realistic outputs.

- To see the vision, you must script it. Visualize and then write out your vision. Written commitments increase follow-through.

YOUR AI-POWERED ACTIONS

ASK:

What is the vision I have for this one beautiful life? How might I use AI to script it out into vivid detail to enhance the probability of success?

APPLY:

Use the Five Steps to a Perfect Prompt to architect your vision with two or three different paths to get there.

AMPLIFY:

Take your vision and co-create a 12-month plan. Every month, check in with AI to create accountability and ask AI to summarize your progress, surface blockers, and propose adjustments.

CHAPTER 3

MAP, MOBILIZE, MOVE UP

Now comes the part I love, where clarity turns into career strategy. In this chapter, we'll take that confidence and put it to work. You'll learn how to use AI not just to dream about what's next, but to design it. You'll see how AI can help you map out your next promotion, position your value, identify growth opportunities, and build a data-driven strategy for your next leap. This is where ambition meets action—where you stop waiting for the next opportunity and start engineering it.

We begin with promotions because advancement is the hinge point of everything else. Promotions determine pay, influence, visibility, and access to opportunity. When women stall here, every other gap widens. And when we move up, everything downstream starts to shift.

Promotions are not informal recognitions of effort. They are institutional decisions filtered through performance ratings, calibration meetings, compensation models, and increasingly AI-enabled evaluation tools. If those systems are biased or opaque, individual excellence alone will never overcome them.

The Promotion Gap: The Myth of Meritocracy

Let's start with what the numbers make painfully clear: the system is not neutral. Women are excelling academically at rates we've never seen before. Women today represent nearly 60% of all bachelor's degree recipients in the United States, a figure that has steadily risen since the mid-1990s.[51] Across the OECD (Organisation for Economic Co-operation and Development, a group of 38 high-income developed countries like the U.S., Japan, Canada, UK, France, etc.), 52% of young women hold higher education degrees, compared to just 39% of young men.

So, let's be clear. We are not just showing up; we are, in fact, over-delivering. We have entered the workforce with gold-star credentials and big-time energy. The ambition is real, the receipts are real, and the climb that we face isn't what we were promised. Academic excellence has not, and does not, translate to equal advancement.

According to the 2024 McKinsey and Lean In Women in the Workplace report, for every 100 men promoted to manager, only 87 women move up, and for women of color, that number drops to 73.[52] This bottleneck occurs at the very first step into management, creating what has been coined as the "broken rung" that compounds throughout women's careers.

We cannot talk about smashing the glass ceiling, the patriarchy, equal opportunities, or any of the gaps if we can't even get past the first rung of the corporate ladder, especially when evaluation systems are becoming more automated and less transparent.

By the time leadership roles are within reach, and the disparity is no longer about ambition, it's about access. Women currently hold only 29.4% of the entry-level STEM roles but just 12.4% of the C-suite positions.[53] The pipeline doesn't look like this because women are unqualified, but because we are underpromoted. This gap has cascading consequences.

Research from *Harvard Business Review* shows that internal mobility, the ability to move laterally or upward within an organization, is a key determinant of long-term retention.[54] Yet 53% of women in tech report it's harder to gain senior-level promotions in their environments than for their male peers.[55]

The myth of meritocracy, the idea of "If I work hard enough, they'll notice me," still runs deep. But when evaluation systems increasingly rely on algorithmic screening, the myth becomes even more dangerous.

This isn't the first time institutions have claimed neutrality while redesigning how advancement works. In nineteenth-century Britain, civil service reforms replaced patronage with competitive examinations designed to create a "merit-based" bureaucracy. Advancement was no longer supposed to depend on connections, but on standardized testing. On paper, this was a triumph of fairness, but in practice, access to elite education and exam preparation favored those who were already positioned within privileged networks. Merit was measured, but the measurement reflected the existing advantage.[56]

The lesson wasn't that merit systems fail. It was that merit systems were designed. Promotion mechanisms always encode assumptions about what counts. Today's algorithmic screening tools are simply the digital descendants of those earlier examination systems.

Recent large-scale experiments show that leading large language models don't evaluate candidates uniformly. In controlled hiring simulations, female applicants received slightly higher average scores than male applicants with ideal credentials, but Black male candidates were penalized, losing roughly 0.30 points compared to the others.[57] On the surface, this might look like progress for women, but this is still not equity; it's just redistribution. AI doesn't remove the bias; it reshapes it, shifting pressure points and redistributing advantages and disadvantages in a way that can be even harder to detect when systems optimize for pattern recognition rather than fairness. Individuals with overlapping identities become collateral damage.

So, when women appear to gain a fractional edge in one dataset, the broader structure remains unstable because bias that moves is still bias. And without deliberate oversight, automated evaluation layers can quietly harden disparities rather than dissolve them, which means advancement systems now evolve faster than their accountability mechanisms.

"We cannot talk about smashing the glass ceiling, the patriarchy, equal opportunities, or any of the gaps if we can't even get past the first rung of the corporate ladder."

In systems shaped by both human judgment and algorithmic filtering, performance does not automatically convert into opportunity. Visibility, positioning, and strategic clarity determine whether impact travels beyond your immediate sphere. When advancement mechanisms are increasingly mediated by data pattern recognition and digital screening, waiting to be "noticed" becomes a passive (and failing) strategy. If advancement systems are designed, then strategy must be designed as well. That's where mapping comes in.

MAP: Designing Your Route Before It's Defined for You

Mapping isn't about predicting your future but about preparing for it. It's about taking the raw material of your ambition and turning it into architecture.

For generations, women have been told that hard work speaks for itself, that excellence is self-evident. Yet the data and daily experience of women show the opposite. Careers don't unfold like bureaucratic staircases. Rather, they branch, they loop, they move sideways, they move backwards, and often, for a long time for women, they stall.

However, opportunity favors those who can see around the corners before others do. Career mapping gives you that vision. But career mapping with AI takes your possibilities to places you would never imagine alone.

Mapping is the practice of stepping back from day-to-day work to intentionally understand the landscape you're operating in. It's how you stop navigating your career by instinct alone and start seeing the bigger picture, clearly. In practice, mapping is both an act of the imagination and one of evidence. It requires the same discipline companies use when they design their five-year strategic plans, trend analysis, scenario modeling, and measurable milestones, but applied to the scale of a single person. Mapping asks you to look beyond your current role and performance and consider questions like:

- What patterns are emerging in my industry?
- What skills are becoming obsolete?
- Who controls the levers of visibility where I work?

The answers are rarely static.

The McKinsey and Lean In 2024 Women in the Workplace report shows that women are exiting middle management at the highest rate in a decade, often citing "invisible ceilings" and a lack of sponsorship.[58] At the same time, the World Economic Forum forecasts that over 40% of current job skills will change by 2030, with many driven by automation and AI integration.[59] Taken together, these shifts make passive career planning a death sentence. The ladder is already unstable, roles are shifting faster than career paths can be paved, and the rules are changing in real time. Mapping is how you respond. It reframes uncertainty as agency by helping you see where change is happening, where opportunity is opening, and where to position yourself next. In other words, when the rules are changing, continuing to operate the same way is a risk. Learning how change is unfolding, and positioning yourself accordingly is the real advantage.

In traditional corporate structures, advancement tends to be linear—a ladder whose rungs are visible and prescribed. However, today the modern career looks more like a metro system in New York City at rush hour: multiple tracks, unpredictable transfers, and a few bottlenecks that you can't see until you're standing in them. The leaders who do well in the crazy metro system are the ones who study it; they anticipate issues like congestion and hop to another line before the chaotic crowd forms.

When the map becomes too complex to read alone, you need a system that can process it at scale. And this is where AI enters the picture.

Large-scale data systems can now scan labor market signals, leadership competency models, and emerging scale taxonomies to help you spot inflection points long before they reach your job description. When used internally, AI can become a partner that visualizes possibilities you may overlook.

Just five years ago, when I coached women on their next career move, the process was entirely manual. We'd start with a gap analysis, looking at where she was now, finding a role one or two levels above her, and dissecting the job description line by line. Together, we'd identify the responsibilities, experiences, and skills she still needed, and then build a plan to close those gaps.

For example, if I were working with an HR generalist who wanted to move into an HR Business Partner role, I would send her to LinkedIn (or any job board) and have her collect half a dozen postings. We'd go through each one, highlighting the common threads and manually comparing them to her resume. Then, we'd design a plan of attack.

With AI, this process looks nothing like it used to.

If you were using the power of AI to map your career, you could feed in job postings from your field or industry, performance criteria from target roles, leadership frameworks from your organization or your desired organization, and ask what competencies are common across these. What am I already strong in, and where am I underinvested?

You could also ask AI to examine your current resume and job description and compare them with a desired job description to help you see things that you might not have seen the first time. You could ask AI to identify the skills and qualifications you already have that may not translate directly, and then help you rewrite your resume to better match the desired job description.

Think of AI as a system that's continually recalibrating as new data emerges, offering additional visibility and flagging opportunities you may never have known existed.

Career mapping with AI doesn't eliminate uncertainty, but it *does* translate it into information you can act on. It's a deliberate, iterative process that can transform vague aspirations into a plan that's executable. And for women navigating a system that was never meant with us in mind? This concept becomes a "not-so-secret" weapon that you can use to be more strategic, thoughtful, and intentional about every career move you make.

The Science of Mapping: From Linear to Adaptive Strategy

Career development is increasingly understood in the behavioral sciences as a feedback loop, not a ladder. This, of course, goes totally against what we, as humans, have been trained to believe. We believe we should be advancing with each step of the journey. However, according to recent research, individuals who actively engage in periodic self-assessment and external benchmarking achieve promotions at higher rates than those using ad hoc strategies.[60]

Women work hard and often accumulate credentials, as shown by the aforementioned OECD statistics, but what really matters is mobility, not just qualifications. Research by the McKinsey Global Institute reveals that nearly 80% of the U.S. gender pay gap is attributable to differences in cumulative work experience and mobility patterns.[61] In other words, degrees alone are not sufficient. What matters is how you position them, how you move them, and how you signal that motion to the right decision-makers. In my coaching practice, we look at three critical layers:

The Competency Landscape: Where is your field going in the next one to three years? What skills will become currency in the next three to five years?

Visibility Architecture: Who decides who gets promoted in your organization? What unspoken (or spoken) criteria are used? Where do you currently stand?

Your Strategic Mobility Plan: What lateral moves, project bets, and networks will accelerate your trajectory?

MOBILIZE: Turning Strategy into Movement

You cannot mobilize without a map, and you cannot move up without mobilization. Once you have the map drawn, forward motion becomes the goal.

This is often where women encounter the quietest and most persistent form of friction: recognition moves more slowly around them, and caution shapes how they move within it. Even when women are qualified, competent, and ambitious, the machinery of advancement runs on more than merit; it runs on visibility networks and language, and on the ability to *signal* readiness as clearly as we perform it.

Mobilizing begins by recognizing that readiness must be both demonstrated and documented, and this is where AI can amplify the process. When used intentionally, AI becomes a performance accelerator. It helps women practice, prepare, and position themselves for opportunity instead of waiting on endorsement.

Let me show you what that looks like in practice.

Language Calibration: Drop the Disclaimers, Claim the Credit

In organizations governed by performance systems, language is data. Self-descriptions, performance summaries, and stakeholder

THE CAREER LEAP ACCELERATOR

Here's an example prompt that will help you get started. Note that each layer builds on the previous one.

ROLE: You are Herminia Ibarra, author of *Working Identity* and a global expert in adaptive career design and leadership transitions.
GOAL: Transform my resume into a behavioral Career Mobility Map that highlights not only what I've done, but how I've moved, adapted, and influenced—uncovering hidden experiences that drive my next-level mobility.
CONTEXT: I'm a professional woman uploading my current resume for analysis. I understand that real career progression isn't linear but cyclical—a loop of reflection, repositioning, and reinvention.

I want to identify:

1. Which of my visible experiences are most promotable in my field.
2. What invisible skills or leadership moments don't show up on my resume.
3. Where my next mobility bets (projects, roles, networks) should be to accelerate growth.

CONSTRAINTS: Use a behavioral-science lens (career as feedback loop, not ladder). Prioritize mobility over credentials. Provide actionable recommendations across three layers:

- Competency Landscape (one to three years): Skills that will rise in value.
- Visibility Architecture: Who decides advancement and how to become visible.
- Strategic Mobility Plan: Lateral or project moves that create compounding growth.

CONFIRMATION: Please ask up to three clarifying questions, one at a time, to help you get started.

narratives feed the same evaluation pipelines that influence promotion outcomes.

Language is one of the most underused tools in career advancement. The words we choose, how we describe our work, our results, and ourselves signal credibility long before performance reviews or promotion discussions ever take place. Every sentence we speak or write leaves a trail of perception. It tells people how we see ourselves and how we expect to be seen.

When those narratives enter talent management systems or calibration discussions, they become part of the measurable record. Infrastructure doesn't distinguish between modesty and impact; it records what's documented.

When we default to soft phrasing, like "I helped" or "I supported," or even worse, "I was lucky to," we minimize the ownership and expertise that our work deserves. The shift is subtle but powerful. It's moving from the language of participation to the language of leadership. AI can be a powerful mirror here by analyzing tone and phrasing, and it can help you spot patterns in your writing that may dilute your authority. Additionally, you can upload a performance summary and ask AI to review and assess what you've written. For example:

"Highlight any places where I have used vague or minimizing language. Suggest revisions that emphasize ownership and outcomes."

The goal isn't to sound robotic or overconfident but to sound aligned with your actual level of impact.

Over time, these small linguistic adjustments rewire how others perceive your leadership and, more importantly, how you perceive yourself.

Evidence Building: Bring Your Receipts!

In promotion discussions, potential is persuasive, but what the research shows is that men are promoted based on potential, and women are promoted based on repeated performance.[62] So, until this injustice has been righted, let us be sure that we're capturing the facts and bringing the receipts of our proven potential to the table.

While this sounds like common knowledge, the data shows otherwise. According to *Harvard Business Review*, women reference quantitative outcomes in self-evaluations 24% less frequently than men, a pattern linked to both socialization and bias in feedback systems.[63] Here, AI acts as your evidence collector. Retrieval-augmented systems can parse emails, project trackers, and reports to compile measurable outcomes, essentially generating a living portfolio of your impact and your leadership. Have you ever struggled to capture everything you've done when it comes time for a performance review? This is how you move from relying on your memory to building a record of impact.

A practical workflow might look like this:

ROLE: Act as a professional development analyst who specializes in promotion cases inside large global technology companies. You understand how leadership evaluates scope, impact, and trajectory, not just new activity.

GOAL: Produce a one-page executive-ready summary of my measurable business contributions from the past 12 months that clearly support a promotion decision.

CONTEXT: I manage a hybrid team in a global tech company. Key priorities include innovation, efficiency, and retention. This summary will be reviewed by senior leadership who may not work closely with me day-to-day.

CONSTRAINTS: Use only verifiable metrics from project documents, OKRs, and performance dashboards. Present on one page. No speculation, fluff, or soft claims. Tie outcomes to business impact, not effort.

Keep the final output to one page and write in a clear, consistent, promotion-caliber tone.

CONFIRMATION: Before summarizing, ask up to five clarifying questions, one at a time, to accurately assess the scope of the impact, scale (including team, region, budget, or reach), the relative difficulty or complexity, and what leadership values most at the next level. Do not generate the summary until the interview is complete.

When you can outline your story with real data, you can start to map out your future with a deep sense of clarity. But if you're going to tell your story with conviction, a flawless execution demands rehearsal. That's where scenario planning comes in. While you won't be able to plan for every possibility, you will be able to plan (and prepare) *for most.*

Scenario Planning: Practice Without Panic

In the keynote speaking world, we often talk about "doing the reps." That's a dressed-up way of saying that you have to practice. You have to rehearse. You need to do the dang thing time and time again before you get really good at it. And as much as we hate rehearsing, it's what makes a boring presentation transform into a standing ovation keynote.

But we HATE rehearsing. I personally put it off until the last possible moment. When I did the deep work to figure out why I had such an aversion to rehearsing, I made the discovery that it was because I hate being vulnerable and I hate messing up and looking stupid. But rehearsing is one of those things where you have to be bad at it, over and over again, before you can be great at it. Again, you have to do the reps.

AI's simulation capability allows you to get those "reps" when it matters most. For example, an AI model can replicate a typical performance review conversation using historical company data, common objections, and behavioral economics insights. The tool can rate

"When you can outline your story with real data, you can start to map out your future with a deep sense of clarity. But if you're going to tell your story with conviction, a flawless execution demands rehearsal. That's where scenario planning comes in."

responses for clarity, confidence, and alignment with organizational goals that provide constructive feedback. This mirrors cognitive behavioral training, commonly known as CBT.

CBT is a structured, time-bound process that invites you to map the link between what you think, feel, and do, and then deliberately rewires that loop through targeted, repeatable exercises. It helps you identify unhelpful patterns, lean into more adaptive thoughts and behaviors, and build a behavioral muscle for real-life change.[64] That repetition reduces anxiety by taking a situation that's filled with uncertainty into one of familiarity.

This rehearsing to reduce anxiety could be applied in other situations, both professionally and personally. I personally love using AI for tough conversations, such as role-playing the message of a reduction in force, a termination of a long-standing employee due to performance issues, or practicing for a high-stakes interview. It could be used in rehearsing your counter-offer for a coveted new role, or in asking your partner to be more hands-on with the kids. (After all, what conversation could be more high-stakes than offloading kid responsibilities to another parent to free up your energy and time?)

When used consistently, this rehearsing builds *readiness loops*—interactive cycles of rehearsal, reflection, and refinement. Every time you rehearse, you're completing that readiness loop, making you more and more ready for the high-stakes/tough conversation. Each pass through makes the next conversation easier, more effective, and more natural. Each pass through makes you more confident.

Do the reps, and use AI to do the reps with. (No more awkward role-playing or practicing in front of a mirror, I promise!)

Mentorship on Demand: Bridging Access Gaps

Access to sponsorship is one of the most unevenly distributed forms of career capital, and when access is uneven, infrastructure becomes the equalizer. In January 2025, my private community for women, HER

Collective, went live. I wanted the women inside HER Collective to be able to ask me questions, send me messages, and talk to me whenever they wanted. But the truth is, that's just not feasible.

So, I learned how to create my first custom agent in ChatGPT. I named her Sheryl after Sheryl Sandberg from Facebook/Meta. (This was before the book *Careless People* by Sarah Wynn-Williams came out; my view on Sheryl has since shifted, but that's a story for a different day.) My AI Sheryl was created to be my digital twin.

I provided Sheryl the entire manuscript of my book, *Glass Ceilings and Sticky Floors*. I trained her on the frameworks I use, the coaching methodologies I teach, and the mindset tools I live by. I wanted her to not just talk like me, but coach like me. Advise like me. Think like me. Then I embedded Sheryl right inside my community so that my members could access her anytime they needed.

At first, it was an experiment—a way for me to explore what was possible with AI, and for my members to have a little fun. But very quickly, Sheryl became something more. She became a mentor who never sleeps, a coach who never forgets what you've said before, and a thought partner who's always available at 2 p.m. or 2 a.m., always ready to listen. And that matters because mentorship and sponsorship remain the strongest predictors of promotion.

According to the McKinsey & Company and LeanIn.org's Women in the Workplace research, women with active sponsors were 1.5 times more likely to advance to senior leadership within two years than those without.[65] However, access to mentorship remains extremely unequal. Research shows that women are less likely than men to have mentors or sponsors supporting their advancement. The numbers tell the rest of the story. Studies across organizations like LinkedIn, AnitaB.org, and Mentorink suggest that mentorship is linked to higher retention, stronger job satisfaction, faster promotion rates, and increased earning potential over time.

Ladies! This is impacting our wallets! And many women report hesitation or lack of confidence when it comes to asking for mentorship or sponsorship. Guess what? With AI, she doesn't have to. The ladies in HER Collective don't have to come to me. They can go to Sheryl.

Now, let me be clear. I do not believe that AI can REPLACE human mentorship and interaction, nor do I think it should. But it is capable of providing structured guidance when access is limited, and when someone might not be ready to talk with a mentor about something.

Inside HER Collective, one woman shared that she regularly worked with Sheryl to flush out the ideas in her head that she wasn't quite ready to speak out loud—not to me, not to anyone.

And in this case, because I built a custom GPT agent, they are provided with the next best thing to a coaching session with Erica Rooney—access to Sheryl.

This kind of mentorship proxy builds the habit of asking better questions. The output of those better questions becomes a blueprint for structured growth we can follow while we continue to seek real-world mentors and networks.

•

At its core, mobilization is about reframing motion as mastery. It's a deliberate progression supported by systems that make learning continuous. When women begin to use AI as a reflective collaborator, using it to track progress, refine communication, and analyze opportunities, it creates a dynamic record of advancement that's both internal and external. Internal because it strengthens confidence, and external because it produces data that organizations reward and recognize.

This convergence of personal growth and technological augmentation is what defines the modern leader. It's not about adopting every tool. It's about selecting the ones that make your movement measurable, visible, and repeatable.

MOVE UP: From Prepared to Positioned

If career mapping is about clarity and mobilizing is about forward motion, then moving up is about momentum with intention. It's the moment where competence meets communication and where performance must translate into leadership presence.

Women don't move up by accident or luck. They move up by making performance visible to the right stakeholders, at the right cadence, in the right language. Strategic visibility isn't vanity; it's how organizations register value and allocate opportunity. The structural headwinds women face are real.

We've already talked about the broken rung. We know that women are promoted at lower rates, and that the shortfalls compound over time, reducing representation further up the ladder.[66] We know that internal mobility, both lateral and upward movement inside a firm, is one of the strongest predictors of retention and growth, but those movements have to be engineered and signaled.

And guess what? Those internal promotions and mobility rely heavily on whether an employee's work is known beyond their immediate circle.[67] It's not just about what you do, but who sees it and who understands its impact. In the tech setting, 53% of women say it's harder for women to gain senior-level promotions than it is for men, which makes deliberate positioning even more important.[68]

To put it plainly, you can be performing at the top of your game, exceeding expectations, and crushing goals, but if no one can see it? It doesn't matter.

Strategic Visibility: The Currency of Career Growth

One of the common misconceptions I originally had about visibility as a new executive was that I had to be everywhere. I needed to show up in all of the places, all the time, and always be ready. But what I've learned is that it's not about being everywhere; it's about showing up in the right room—the rooms where decisions are made.

So how do you get into those rooms? Well, friend, you have to be visible.

Then, we run into the age-old problem of which comes first: the chicken or the egg? How can I be visible in the right room if I have to be visible to get in the room? And while I will never know if it was the chicken or the egg, I do know one surefire way to be visible is to showcase your wins.

Women often hesitate to share wins because they are worried about how it will be perceived. We've been told for decades that humility is a virtue, that good work will "speak for itself," and that the payoff will follow. But research and reality do not agree on this issue.

There's a double bind in this situation. For many women, particularly women of color, the act of advocating for oneself is layered with social, professional, and emotional risk. Speaking up can amplify your value, but it can also trigger bias. Women who self-promote or negotiate assertively (like a man would) are more likely to be labeled as "aggressive," "abrasive," or "difficult," while men who exhibit the same behaviors are described as "confident" or "decisive".[69] (OK, how hard are you rolling your eyes right now?)

A *Harvard Business Review* analysis captures this dynamic clearly. In "How Women in Leadership Can Shape How Others See Them," researchers Julia Howell and Alice Klein describe how women in power face what they call a well-documented "lose-lose" scenario. Women leaders are often seen as either competent but cold or likable but less effective. This tension forces women to constantly adjust their expression and calibrate warmth and authority in ways that our male counterparts rarely have to consider.[70] Their findings show that while women consistently earn higher ratings for empathy, collaboration, and communication, they are simultaneously penalized when they display the same levels of assertiveness and strategic control that earn men praise.

> **"AI can help women translate quiet competence into visible influence."**

Leadership researcher Alison Fragale calls this balance the art of being a "Likeable Badass," and her research on gender, power, and influence shows that women who rise the fastest are those who learn to blend competence with connection. You are assertive enough to earn authority, but authentic enough to sustain trust.[71] Alison also details in her book, *Likeable Badass*, that people cannot value what they cannot see, and that competence, no matter how strong, remains invisible unless it's communicated. Her research shows that women who strategically share their impact, connect outcomes to organizational goals, and build reputational capital through consistent visibility aren't perceived as self-promoting, but rather as credible.[72]

So how does this fit in with AI? Let me connect the dots. AI can help women translate quiet competence into visible influence. It can surface measurable outcomes buried in emails or project milestones. It can refine language to reflect leadership without distortion or apology. (Don't forget it can act as a "rep" when needing to prepare for a challenging conversation, promotion request, or negotiation!)

Women who learn how to really work with AI can customize an agent to become their personal brand strategist, their PR agent, and their career futurist, all trained to maximize visibility and impact. One of the most useful prompts I used while researching this book was simple: "Based on my current projects, what trends could I lead in this organization or industry over the next 12 months?"

Remember, context is key, so to use this, you have to be regularly working with your AI, or providing it with the relevant context in documents and other supported materials (your goals, wins, reflections, etc.).

Can you start to see how, with the power of AI, you have the ability to engineer your own visibility? Not only can you do this, but you must do this. The future won't reward women who are hard-working or those who carry the invisible labor; it will reward those who are seen, and those who know how to make their brilliance impossible to ignore. But being seen isn't something that just happens. It's something you build, one documented win at a time. And one thing we know, as women, is that excellence doesn't guarantee advancement. In systems where documentation determines advancement, preparation becomes structural. Decision-makers act on what's visible, and what's visible gains momentum over time.

Preparing for the Move-Up

One of my favorite human resource sayings is, "If it isn't documented, it didn't happen." I have no idea who to credit that to, but in real talk, if you don't have the receipts of your success, it didn't happen. The corporate world moves fast. Leaders change, teams shift, and if you aren't careful, the visibility you worked hard for could vanish in an instant if you don't do these four things: document, demonstrate, develop, and have a dialogue.

Document

AI can help you track, quantify, and curate your achievements in real time, so you aren't waiting until the last day of performance review season and trying to remember what you did in January. Think of AI as your PR manager who's constantly gathering your wins and cataloging them. (Bonus points if you ask AI to surface patterns over time, identifying themes across quarters.)

Demonstrate

AI helps you articulate your impact in ways that resonate with decision-makers and cross-functional teams. You can use AI to rehearse your executive presentation by pasting in the outline and asking AI to analyze it for tone and framing. Then, go rock that presentation like the "likeable badass" you are.

Develop

The people who get promoted are the people who never stop learning. Use AI to identify the one skill that would have the highest ROI on your advancement. Then, level up by asking it to build a learning plan using credible sources.

Dialogue

Using AI to simulate real promotion discussions is one of the most effective ways to have these conversations confidently. I can't stress enough how important "doing the reps" is. AI is the perfect partner.

•

Promotion systems are no longer static; they are increasingly digitized, standardized, and shaped by data flows that outlast individual managers. What becomes visible inside those systems builds over time, and what remains undocumented disappears. AI accelerates this dynamic.

Bringing It All Together

You've done the inner work, mapped what's next, and started mobilizing out loud, not just in the privacy of your own home. The next step is simple, but not always easy: make your performance impossible to ignore.

This chapter was never about being louder. It was about being clearer about turning quiet excellence into visible influence. You learn to map with precision, seeing around corners instead of waiting for job postings; to mobilize with intention, documenting outcomes, calibrating language, and building proof; and to move up by designing visibility that travels into rooms you might not be in yet. That's the difference between hoping for an opportunity and engineering an opportunity.

AI is your accelerant. It doesn't change who you are, but it does help others finally see what a "likeable badass" you are. (Thank you, Alison Fragale, for this gem of a term!). It translates your results into a leadership language that can be heard and understood by everyone. It rehearses the moments that once made your stomach drop and heart rate go through the roof, until confidence replaces that fear.

When you use AI in this way, it becomes the quiet infrastructure that helps you climb the corporate ladder. So, here's your operating rhythm going forward:

- **MAP** with data and intention. Use AI to scan the horizon, test scenarios, and target the roles and skills that will matter 12 months from now.
- **MOBILIZE** with receipts. Keep a live trail of your outcomes and successes, and translate them into executive terms. Then, circulate them where decisions are made.
- **MOVE UP** with strategic visibility. Share your wins and align your narrative to the organization's priorities.

If you do these three things consistently, your readiness becomes embedded into institutional memory. In environments governed by data flows and digital review systems, visibility compounds. What's documented influences calibration. What's calibrated influences compensation. What influences compensation shapes long-term economic power. Infrastructure doesn't pause for modesty. It records what's measurable and scales what's documented. Design your visibility accordingly.

TLDR

- Visibility isn't a vanity metric; it's critical to your success. The pipeline leaks not from lack of talent, but from lack of access, sponsorship, and visibility.

- AI can analyze job descriptions, industry trends, and internal frameworks to help you identify which skills are rising in value, where bottlenecks occur, and what lateral or project moves build long-term leverage.

- Women are often promoted on proof, not potential. AI helps you collect receipts in real time, rewrite self-evaluations with stronger ownership language, and rehearse key conversations through simulated "reps" until confidence becomes muscle memory.

- Strategic visibility is the currency of career growth. AI acts as your PR strategist, surfacing measurable outcomes, refining your narrative, and forecasting future influence.

YOUR AI-POWERED ACTIONS

ASK:

What's the next level of my career that I want to reach? And how might I use AI to map the clearest data-backed path to get there?

APPLY:

Feed AI your resume, recent wins, and a coveted target role. Ask it to identify your promotable strengths, hidden skills, and the visibility gaps that you need to close. Then, map out the next three strategic (but clear!) steps you will take.

AMPLIFY:

Build your "visibility vault" inside AI. Track every win, every metric, and every moment of impact. Each quarter, ask AI to summarize your progress, highlight emerging themes, and forecast your next strategic career move.

PART 2

CLOSING THE AUTHORITY GAP

Making Your Power Visible in AI Systems

Awareness changes how you see the system. Authority changes how the system sees you.

In the AI era, advancement is no longer shaped solely by direct managers or informal sponsorship. It's increasingly filtered through systems that rank, surface, summarize, and standardize performance. Algorithms influence recruiting pipelines, and digital profiles shape discoverability. Evaluation tools structure how impact is interpreted. Visibility now operates through the system itself.

This shift creates a new kind of gap. The Authority Gap is the distance between doing strong work and having that work recognized inside systems that rely on signal, documentation, and pattern recognition. Talent still matters. Effort still matters. But visibility now determines how they are distributed.

Closing the Authority Gap requires intention. It means translating quiet competence into visible outcomes. It means documenting impact before the performance review cycle begins. It means understanding how decisions are made, how influence travels, and how digital presence shapes perception long before a conversation happens in a room you are not yet in.

This section moves you from private readiness to public positioning. You will learn how to map opportunity strategically, mobilize your results with clarity, and design visibility that compounds over time. You will strengthen your decision-making under pressure and install rhythms that sustain forward motion rather than relying on intensity alone.

Authority grows from alignment between your contribution and the systems that recognize it. Impact that's clearly seen moves beyond your immediate circle and settles into institutional memory. This is where advancement becomes more intentional, and where the gap between effort and opportunity begins to close.

CHAPTER 4

DIGITAL PRESENCE AND INFLUENCE

For the past decade or so, experts have talked about the importance of a strong digital presence, but in the AI era, it's become increasingly essential for leadership success.

Artificial intelligence is changing how reputations (and brands) are found and perceived online. The old way of being discovered is changing, and fast. You can't rely on word of mouth, organic reach, or a well-timed LinkedIn post anymore. The digital landscape has shifted, and the fact of the matter is that if you aren't digitally legible, you're professionally invisible.

Search engines, AI-driven recruiting tools, internal talent platforms, and even recommendation systems inside LinkedIn are constantly ranking, surfacing, and filtering professionals based on digital signals. Visibility is no longer just about reputation; it's about discoverability inside algorithmic systems. This dynamic isn't new.

In the late nineteenth century, commercial credit reporting agencies began converting informal business reputation into standardized, shareable credit files. Once reputation became record, lenders no longer relied solely on personal networks; they relied on documented profiles. Access to capital followed what had been recorded.[73] Digital visibility functions the same way. Once expertise becomes indexed data, opportunity follows the index.

Algorithms don't "notice" quiet excellence; they index patterns, surface frequency, and reward clarity. Which means that if your work isn't visible in structured, searchable, consistent ways, it will go unnoticed and unindexed.

Women leaders in particular can strategically leverage AI tools to enhance their visibility and influence while addressing the unique challenges they face around self-promotion and consistency. In this chapter, we'll explore four key areas: building credible thought leadership, sharpening personal branding, maintaining consistent visibility without burnout, and overcoming emotional barriers to being seen. But before we dive in, let's talk about the importance of digital visibility and the impact it can have on your career.

The Career Impact of Digital Visibility

A strong online presence can significantly influence career opportunities. The data shows that over 95% of recruiters use LinkedIn regularly to discover and evaluate talent.[74] In fact, an estimated 122 million people have landed an interview through LinkedIn, with 35.5 million actually getting hired through connections made on the platform.[75]

For women seeking leadership roles, these statistics highlight the importance of being visible online. Research shows that 65% of professionals consider an online impression as impactful as an in-person one.[76] By sharing content, engaging in industry discussions, and showcasing expertise on different social platforms, women can attract recruiters' interest and signal their leadership readiness.

Research also suggests that strategic digital visibility, not volume, correlates with career mobility and recognition. Studies published by *Harvard Business Review* show that employees who are more visible for their expertise are more likely to be perceived as high potential, invited into stretch opportunities, and considered for advancement, even when performance is less comparable.[77]

But in an AI-shaped landscape, visibility carries even more weight. Your digital footprint increasingly functions as data. The content you produce, the ideas you engage with, and the topics you are associated with all contribute to how systems categorize and surface your work. If you're not actively shaping that narrative, it will still be shaped, just not by you. Make no mistake about that!

And here's where this conversation becomes more uncomfortable for many women. For decades, we've been socialized to believe that humility is safety, that excellence should speak for itself, and that self-promotion carries social risk. In human-centered environments, that caution sometimes protected us. In algorithmic environments, it simply makes us invisible. Digital systems do not reward restraint. They reward clarity, repetition, and recognizable patterns. If women hesitate to publish until something feels perfect or to speak until they feel fully ready, the cost is no longer just emotional. It becomes structural. Silence does not read as humility to an algorithm; it reads as absence.

On platforms like LinkedIn, visibility does not come from posting constantly, but by contributing ideas that are clear, distinctive, and aligned with your expertise. In an environment saturated with content, what stands out is not frequency, but positioning.

While only a small percentage of users post regularly, the majority of content receives limited engagement, which means most professionals are not competing for attention so much as failing to communicate what they are known for. Visibility that drives career mobility is not about being seen by everyone, but rather being *recognized by the right people* for the right work.

When we look at it through this lens, thought leadership functions less like broadcasting through a megaphone and more like pattern reinforcement. When your ideas, results, and perspective are consistently legible (across conversations, across documents, and digital spaces), you reduce ambiguity about your value. That clarity is what enables recognition, sponsorship, and movement.

"If you're not actively shaping that narrative, it will still be shaped, just not by you. Make no mistake about that!"

What's most powerful about an online presence is that women can also control their professional narrative. A personal brand crafted through thoughtful posts and articles allows you to highlight the skills and achievements that might otherwise be overlooked. One leadership advisor, as quoted in *Forbes*, said that establishing a clear personal brand "allows women to increase their visibility, control their narrative, combat unconscious bias, and open doors to new career opportunities."[78]

In a world where women's advancement is an uphill battle, strategic digital presence is the differentiator. It ensures that your name comes up in the right conversations, whether that's a recruiter searching for candidates or a senior leader seeking an internal collaborator. In fact, hiring managers increasingly expect to find some sort of online footprint. In one survey, one in five employers said they are unlikely to consider a candidate with no social media presence at all.[79]

The message is clear: visibility matters. The woman who shares her insights online signals expertise and demonstrates confidence and communication skills. These are qualities associated with leadership. So why does building a digital presence feel overwhelming?

Building Credible Thought Leadership with AI

Digital platforms are not neutral stages. They are ranking infrastructures. Every post, profile feed, and interaction becomes structured input.

Let's be real for a moment. It can be exhausting having to show up in the digital world today. The pull to be constantly creative and "on-brand" on all platforms can become a serious energy drain that derails your climb.

The system keeps moving whether you show up strategically or not. It continues to index your work, your presence, and your visibility over time. I've felt this pull while growing my own brand. What has changed is that with the power of AI, showing up strategically no longer requires the same level of effort it once did.

One of the most powerful moves you can make today is to let AI act as your most energized, highly organized, and never-tired personal assistant who can amplify your thought leadership from brainstorming ideas to perfecting a final draft. Tools like ChatGPT, Google Gemini, Microsoft Copilot, and Claude all serve as very smart, capable thought leadership partners that can be trained over time, allowing you to multiply your output without adding hours to your work week. I personally use AI at every stage of content development—research and brainstorming, drafting and editing, tone matching, and strategically scaling output—to get more done in less time, while maintaining the integrity of my ideas, my voice, and my thoughts and opinions.

The main takeaway is this: AI works best as an enhancer, not as a replacement. It can refine your ideas and save you time, but your unique perspective must be at the forefront. As one *Inc.* article put it, simply mass-producing ChatGPT-written LinkedIn posts is brand-damaging and will be filtered out by algorithms that prioritize originality. The goal is to use AI to sharpen your messages and get them out there while ensuring the ideas and insights are authentically yours.[80]

AI-Powered Personal Branding Strategy

Building a personal brand in the digital age is another area where AI tools can give women leaders an edge. From optimizing bios and profiles to monitoring online reputation, AI can help craft a consistent, compelling strategic narrative. That narrative becomes structured input inside discovery systems.

USING AI FOR CONTENT DEVELOPMENT

Research and Brainstorming: Thought leadership thrives on original ideas and insights that are backed by data. While AI cannot be the originator of those original ideas, it can be a powerful research partner, helping you to sift through information, articles, and relevant stats and trends. It can even analyze audience conversations and customer sentiment, helping you to zero in on timely, high-impact ideas for articles, social posts, or keynotes.

Drafting and Editing: If you've ever had an idea but struggled to get that first sentence going, generative AI excels at overcoming the blank page and refining rough drafts. You can use AI to help you generate outlines and link concepts together, which will help you kick-start the writing process.

On the flip side, once you've written a draft that needs polishing, you can use AI to suggest clearer wording, tighten the text, or adjust the tone. (All of which I am doing while writing this book!) And I'm not the first. LinkedIn co-founder Reid Hoffman famously co-wrote an entire book with the aid of an AI writing model, demonstrating how AI can enhance content creation when guided by strong human ideas.[81]

Tone Matching: Maintaining an authentic voice is crucial for credible thought leadership. Once you've gotten familiar with AI, you can even train AI to mirror your tone. Over time, a personal AI chatbot can learn to think and talk in a style that matches you, which is useful for drafting content that still sounds like you, even if the AI took the first pass.

When I first started working with ChatGPT, I would often prompt the chat by telling it my tone was a blend of Mel Robbins and Jenna Kutcher—empowering, educational, direct, and research-heavy. Over time, I've continued to feed my AI transcripts from past talks or podcast interviews so that it can learn to better match my voice and tone.

Scaling Output (Strategically): AI makes it easier to consistently produce content like blog posts, LinkedIn updates, and newsletters, which is important because thought leadership rewards both quality and consistency. When you automate lower-level tasks and speed up writing, AI helps free women leaders to share insights more frequently.

But let me offer you a word of caution: you cannot blindly turn out AI-generated posts without human oversight. Experts warn that flooding the internet with generic machine-written content is not effective and can even be harmful to your brand since it lacks your genuine expertise that true thought leadership demands.[82] Not only that, but it can backfire against findability since it lacks actual thought leadership.[83]

Profile and Bio Optimization

There isn't a single profile or bio of mine that exists that hasn't been optimized without the power of AI. You can easily enhance your bio by asking a tool like ChatGPT to rewrite your bio in a more dynamic way, highlighting your achievements in an engaging yet authentic tone. I have found it useful to iterate through several AI-suggested versions and then pick the elements that best capture my voice and what I want to convey. You can use the prompt below to get started.

Identifying Brand Themes

Think about the themes that define your work, the ideas you keep circling back to, the stories you can't stop telling, and the conversations you want to be known for online. Those are the threads that make up your digital footprint. And if you aren't sure what yours are, here's where AI can help. Instead of guessing, ask it to analyze keynotes, social posts, or even podcast transcripts if you have them. It can quickly surface patterns, the words, emotions, and topics that keep showing up in your work, so you can see what your brand is already saying.

If you're already clear on your brand themes, you can use AI to amplify them. Have it study your content through a new lens and pitch fresh angles that deepen your credibility and reach:

- "Analyze these samples and list my top five themes with one sentence that describes each."
- "Pull phrases and emotions I use most. What feelings do I consistently create?"
- "Which stories land as credibility builders, and why?"
- "Give me three new angles for each theme that feel current and useful for my target audience."

A PROMPT FOR OPTIMIZING YOUR PROFILE AND BIO WITH AI

Use this prompt when your current bio feels flat, overly corporate, or disconnected from the depth of who you actually are. It's designed for moments when you're evolving, repositioning, or stepping into a bigger stage and need your profile to reflect not just what you've done, but the transformation you create and the presence you carry.

ROLE: You're a world-class interviewer and brand strategist who specializes in helping thought leaders articulate their unique genius with clarity, warmth, and power.

GOAL: To interview me and extract the stories, proof points, and emotional through lines that will help craft a magnetic, credibility-rich professional bio and LinkedIn profile that feels 100% *me*—authentic, human, and high-impact.

CONTEXT: I'm an experienced speaker, author, and executive coach building a personal brand at the intersection of leadership, authenticity, and AI confidence. I want my bio to capture both my strategic credibility and my intuitive EQ—not corporate jargon. Think: magnetic, emotionally intelligent, and rooted in impact. You're here to help me surface the transformation I create, the values that drive me, and the essence that makes my presence unforgettable.

CONSTRAINTS: Keep the tone conversational but high-level, like two smart women mapping a legacy brand over coffee. Avoid buzzwords like "powerhouse" or "trailblazer." Keep it real, relatable, and human.

CONFIRMATION: Ask one question at a time to pull out the depth behind my achievements and stories. No generic "tell me about yourself" questions—dig for texture, turning points, and truth. Each question should help uncover one of these: credibility markers, emotional DNA, brand voice, or transformation story.

Visual Branding Aids

On the branding side, AI tools can help create consistent personal brand aesthetics without needing a graphic designer. I use Canva's AI features to generate on-brand graphics and social media templates for all of my social posts and slide decks so that they have a cohesive look that is aligned with my brand colors and style.

Always remember that AI is here to amplify your voice, not to fabricate it. AI should never replace your unique perspective or your authenticity. And I implore you to use these tools to highlight your value, but keep your personal voice at the core.

Maintaining Consistent Visibility without Burnout

One of the biggest challenges I had when starting my business was building an online presence across all of the platforms and sustaining it. I knew the importance of consistently sharing insights, updates, and engaging with my network, but it was draining my energy and quickly becoming exhausting. And if you're an entrepreneur like me, you may be managing a personal brand and your company's brand, which can quickly become A LOT.

AI offers you a way to maintain strategic visibility while lightening the load (and the anxiety!), helping to avoid the overwhelm and burnout that comes from feeling that you must be posting constantly or be online 24/7.

Here are some ways AI can enable consistency with less effort and stress.

Content Scheduling and Automation

AI-powered social media management tools ensure your content goes out regularly, even when you're busy or taking a break. I use the

AI-driven platform Metricool to auto-post on my behalf and determine optimal posting times. This means I can batch my content in advance and let AI handle the rest. When you do this, you maintain a steady presence without the pressure of manually logging in to every platform to post every day.

AI-Assisted Content Creation in Batches

I've found that I'm more likely to post on a consistent basis when I batch-create my content. I will spend one hour a week with an AI tool brainstorming content, organizing my ideas, and mapping out one month's worth of content. In the past, this would have been a longer, more painful process. Now I can schedule a queue of posts to go live when I'm on the beach in Mexico, when inspiration is low, or when I just want a break from social media.

But batching with AI does not mean copying and pasting whatever it generates. AI gives me structure, prompts angles I might not have considered, and helps me see patterns in my ideas. I still rewrite, refine, and infuse every post with my lived experience, voice, and point of view. The goal isn't automation for its own sake. It's amplification.

When used well, AI becomes a creative collaborator. It helps you think faster and more strategically, but your perspective, your stories, and your emotional texture are what make the content resonate. Batch creation works because it removes friction, reduces decision fatigue, and protects your creative energy so you can show up consistently without draining yourself.

Repurposing and Transforming Content

AI can alleviate workload by repackaging your existing content that has performed well into new formats. For instance, if you write a thoughtful blog post, an AI service can summarize it into a series of social media blurbs, extract quote graphics, and generate a short video script from it. This extends the life of one piece of content across multiple channels without requiring you to start from scratch every time. When you automate the grunt work of repurposing content, AI ensures that you always have something to post with minimal extra effort.

Monitoring What Works

Another burnout-inducing aspect of visibility is trying to keep up with the fast pace of online trends and audience feedback. AI analytics can simplify this by automatically monitoring which posts perform well, what times your followers engage, or what topics are trending in your field. Instead of manually checking your statistics or doing deep research on what's working and what isn't, you can let your AI summary tell you. This way, you can focus your efforts on content that is making a difference.

Essentially, AI can help you do less, more reliably. You do not have to be posting in real time or everywhere all at once to grow your influence. When you build a system that uses AI to handle routine tasks like scheduling, idea generation, and cross-posting, women leaders can show up consistently on their own terms.

As one social media expert put it, "Success came when she stopped measuring herself by how much she posted and started focusing on how repeatable and sustainable her process was."[84] AI is a key part of making the process sustainable, so you can maintain a visible presence without running yourself into the ground.

Overcoming Emotional Barriers to Being Seen

Perhaps the most profound gift AI tools offer for women in leadership is how they support the emotional and psychological barriers that hold us back from being visible.

The truth is, so many women carry fears around self-promotion and public presence. We worry about being too much. We're told not to

"When you build a system that uses AI to handle routine tasks like scheduling, idea generation, and cross-posting, women leaders can show up consistently on their own terms."

brag, not to overshare, and not to be flashy, and the perfectionism and imposter syndrome pile up. Research confirms that women are more likely than men to be perfectionists, especially in corporate environments, and that perfectionist drive often keeps us from stretch roles in public visibility.[85]

We learn early that we must prepare until we're flawless. Then you can show up. But that "wait until it's perfect" narrative is a trap. Perfection doesn't exist—and while you are waiting for it to show up, you know what else is on hold? Your visibility.

That's where AI can quietly change the game for you. It isn't a cure for perfectionism or a shortcut past fear. It won't silence the voice in your head that asks, "Who am I to say this?" But it can help you take the next small brave step forward.

AI gives you a safe space to practice, to draft, to explore your ideas without the pressure of an audience watching. You can test your message, find your voice, and see how your words land before you ever hit publish. Over time, that quiet practice builds your confidence. You can start to see the patterns in your work, the themes that matter most, and the strength in your own perspective. You stop waiting for perfect because you realize that clarity grows from taking action.

AI won't do the work for you, but it does lower the barrier to starting. It makes it easier to draft, to refine, to try again. When we stop waiting to be perfect and allow ourselves to be visible in progress, momentum begins.

But that momentum no longer lives only in your confidence. It lives inside systems.

Digital presence now feeds into automated evaluation environments—recruiting filters, internal talent marketplaces, recommendation engines, and search indexing systems. These systems don't interpret intention. They register patterns. Once registered, those patterns influence access.

AI will not generate authority for you, but it will classify what you make visible. Digital systems continuously index expertise, themes, and participation, organizing around what they can see and measure. Advancement follows those signals, and anything outside of them rarely gets considered.

Build accordingly.

TLDR

- Visibility today isn't optional; it's leadership currency. In the AI era, your digital footprint speaks for you long before you enter the room.

- Building visibility starts with clarity. Knowing your core themes, values, and message, AI can help you uncover and refine those patterns hidden in your work.

- For women, the real challenge often isn't skill but self-permission. Perfectionism, fear of judgment, and old social scripts make being "seen" feel risky.

- The goal of using AI is to sound more like you. AI simply lowers the friction between intention and action, so you can lead without burning out.

YOUR AI-POWERED ACTIONS

ASK:

Where am I still waiting to be "ready" before showing up? What part of my work, voice, or story have I kept behind the curtain because I'm afraid of how it might be perceived?

APPLY:

Choose one piece of work, a bio, a social media post, or an article, and run it through AI. Ask it to highlight your recurring themes, the emotions your words evoke, and how your message could land stronger.

AMPLIFY:

Take one of your ideas that you have been keeping quiet about and share it, even if it's messy. The courage to show up imperfectly trains both you and the algorithm to recognize your voice. Over time, consistency becomes credibility, and credibility becomes influence.

CHAPTER 5

DECISIVE POWER MOVES

At the height of my career, I finally had what I thought I always wanted: a seat at the table. The title, the access, the proximity to power, the dollars! But every Saturday night (yes, Saturday), the Sunday scaries would creep in, and I found myself asking, "Is this what I really worked so hard for?"

I was constantly burdened with decisions that had to be made; I was expected to weigh in on all of them, and I was also responsible for outcomes that never felt fully mine. What I didn't understand then is what I know now: power isn't the seat itself. Power is knowing how to use that seat to open the right conversations, make deliberate decisions, and close those decisions in a way that moves you forward. This chapter is about that shift. About how women can stop carrying all the decisions alone and start making them with clarity, alignment, and follow-through.

Let me tell you a story.

When the company I worked at was acquired in a merger with a global tech machine, I knew that my role would be impacted at some point, and that the likelihood of my exit was probable within the next few years. I was honestly surprised that, once the retention bonuses cleared, the executive team was still operating in the same capacity. But as time went on, it became clear that my core values and priorities didn't match the vision of the new leadership structure. I didn't know exactly what I wanted to do, but I knew staying wasn't an option if I wanted to protect my mental health and peace. That's when I leaned into AI to help me create and launch HER Collective.

With the power of ChatGPT, I co-created ideal client avatars, perfected the mission and vision, and outlined the core values that would anchor the business. While building this, a new vision for myself began to take shape. But alongside that excitement came hesitation and self-doubt.

At this point in my career, I had been in corporate America for over 15 years, and I didn't know anything else. I was earning well into six figures, with a strong base salary and a bonus structure that rewarded performance generously. From the outside, it looked secure and successful. But while the role was mentally taxing, it was no longer stretching me. I could do it in my sleep, and that was the problem. I wasn't growing. I felt stuck.

"Power isn't the seat itself. Power is knowing how to use that seat to open the right conversations, make deliberate decisions, and close those decisions in a way that moves you forward."

While I knew I couldn't stay, I didn't know what to do. So, I went to AI. I did deep research on how to start a business. I gathered data. I journaled insights into how I was feeling. I also talked (A LOT) to mentors and people who had "been there, done that." I used logic and instinct, and a heck of a lot of data from ChatGPT. I was trying to figure out: do I stay or do I go? If I stay, I'll be compromising my values, but if I go, I'll be walking away from a lucrative career, the C-level title, those beautiful (but painfully tight) golden handcuffs, and what we all think of as "job security." I needed to make this big decision, but I knew I couldn't, and shouldn't, do it alone.

I call that point in time my decision-making era because I was having to make what felt like MILLIONS of decisions each day. There were decisions about my job, my kids (will this impact them?), my family (will I be able to support them?), my team (what would happen to them if I left?), my clients (where would I find them?). There were decisions about EVERYTHING.

Leaders don't freeze because they can't make a decision, but because they spend their whole day, and all their mental energy, making thousands of tiny decisions. I thought I was overwhelmed. What

I was actually experiencing was the cost of carrying too many decisions without a system to hold them.

Decision Fatigue in Leadership

Women in top executive roles face a staggering volume of decisions daily. Studies estimate that leaders make between 33,000 and 35,000 decisions per day, from trivial choices to big, high-stakes calls.[86] This constant decision load can drain mental resources, leading to decision fatigue: a state where the quality of decisions deteriorates as the day wears on.[87]

And because we're making upwards of 35,000 decisions a day, we're running on what I call "decision overload." Decision overload is when you come home from a chaotic day at work and your partner asks you, "What should we eat for dinner?" and you go into a full-on rage, ranting, "Why do I always have to make the decisions around here!" In reality, your partner was just trying to collaborate with you on a meal plan. But we're exhausted by the mental bandwidth that it takes to constantly choose, manage, and recalibrate, again, and again, and again.

For women in leadership, decision overload isn't just exhausting; it's risky. Research shows that female leaders are evaluated more critically and subjected to harsher scrutiny than their male counterparts, even when performance is equivalent.[88] In systems shaped by gender bias, depleted decision capacity collides with heightened visibility, making each choice more consequential.

Also, that 33,000–35,000 decision estimate assumes a narrow definition of "work." For many women, I would argue that this number is higher. Beyond paid leadership roles, women disproportionately shoulder the second and third shifts: the unpaid labor of running a household and managing emotional and logistical demands outside of work. Think, school supply shopping, planning meals, emotional caretaking, and "all the things" that are required to keep the home running.[89]

The third shift, which was introduced in a 2020 *Psychology Today* piece, describes the always-on work of self-management.[90] It's the pressure to optimize everything, from your body to your mindset to your brand and your business. It's the constant mental churn of tracking goals, gratitude, health metrics from the latest wearable device, and self-improvement checklists, all while trying not to drop any of the balls we're juggling. This is the emotional and cognitive labor of personal growth. And all three shifts, filled with thousands of decisions, are a drain on our mental energy.

These findings show that decision fatigue and pressure aren't just problems that individuals suffer from but structural issues impacting women in leadership. High-level women often navigate a double bind of heavy workplace decisions and disproportionate outside-of-work duties, making effective energy management and organizational support critical to making good, unbiased decisions. So, what makes a "good" decision, anyway?

Decision Quality vs. Outcomes: Overcoming "Resulting Bias"

One of the most common (and costly) mistakes we can make in leadership is confusing decision quality with decision outcomes. They are not the same thing.

Decision quality refers to how a decision is made. The rigor of thinking, the information considered, the assumptions tested, and the processes used. A high-quality decision can still lead to a poor outcome because of factors outside anyone's control. And a low-quality decision can occasionally appear successful through luck alone.

Outcome bias is the tendency to judge the quality of a decision based solely on how it turned out, rather than on the soundness or the reasoning behind it. This bias is well-documented in behavioral science. Research by Jonathan Baron and John Hershey found that people consistently rate decisions with positive outcomes as better, even when they were made using flawed logic, and judge well-reasoned decisions more harshly when outcomes are negative.[91]

In high-pressure leadership environments, outcome bias becomes especially dangerous. It can mask weak decision processes when results happen to be favorable, and it can unfairly penalize thoughtful, disciplined decision-making when circumstances shift or variables change. Research suggests this dynamic is not gender-neutral. Studies show that women leaders are more likely to be evaluated on outcomes rather than on the decision process, reinforcing heightened scrutiny and narrower margins for error.[92]

The implication isn't that women make "worse" decisions (in my humble opinion, we often make the best ones). It's that decision quality matters more when outcomes are judged more harshly. That makes clarity, rigor, and repeatable decision frameworks non-negotiable. When scrutiny is uneven, decision quality becomes a form of protection. A clear process leaves less room for bias to operate.

We've all seen it. The CEO who makes a reckless gamble that succeeds by pure luck and gets praised for his intuition and expertise, even though by all rational measures it was a terrible bet. Over time, this bias impedes learning, and organizations end up over-crediting luck and under-crediting sound decision processes. And because good choices can fail and bad choices can succeed, judging decisions by their results alone is misleading and, honestly, detrimental in the long term.

We should be encouraging decisions that are grounded in reason, informed by research, and supported by real data and insight. Anything less isn't strategic; it's reactive. When we hold this standard, we help leaders move beyond quick fixes and stay anchored in long-term, values-driven decision-making. This is critical because poor decisions are expensive.

The Cost of Poor Decisions and Missed Opportunities

The stakes for strategic decisions are high, and when decision quality declines, the cost to organizations can compound quickly. A 2025 mixed methods study examining the total cost of difficult leaders found that a single ineffective or toxic leader can cost a company between 1.75 and 10 times their annual salary, with an average impact of 8.7 times compensation.[93]

It's important to be precise here. Not all difficult or toxic leaders are poor strategic decision-makers. Some deliver strong short-term results while eroding teams through interpersonal behavior, unpredictability, or control. But even when financial outcomes appear positive, leadership decisions still shape the systems people operate within, including workload distribution, psychological safety, and whether teams can sustain performance over time.

Turnover is one place where those downstream effects become visible. Research shows that 33% of employees have left a role specifically because of a bad manager or toxic leadership environment.[94] That departure is rarely caused by a single decision. More often it reflects the cumulative impact of leadership choices about priorities, communication, and how pressure is handled. The cost of replacing leaders and managers is significant. Gallup estimates replacement costs can reach roughly 200% of a manager's salary.

I'm part of that 33%. I didn't leave because of one bad call but because decision-making dysfunction became the norm. Over time the inconsistency, rework, and cognitive load made the environment unbearable and unsustainable. When experienced leaders exit under those conditions, the organization doesn't just lose headcount, but they're losing institutional knowledge, continuity, and trust. And the people who stay notice, because when strong leaders leave, it signals something deeper is broken.

Even when leaders remain in place, weak decision systems quietly bleed opportunity. A 2024 survey found that nearly half of organizations have stagnant decision-making practices, meaning they fail to systematically review or learn from past decisions, leading to repeated mistakes and missed strategic gains.[95] Capturing the lessons learned isn't enough. If insights sit in documents that no one revisits, the system doesn't learn, it repeats.

The takeaway isn't that every departure or cultural breakdown can be traced to a single poor decision. It's that leadership decisions, especially decisions made under pressure, shape whether organizations intensify value or escalate friction. For women in senior roles where scrutiny is higher and margins for error are almost non-existent, decision quality isn't just a leadership skill, but a form of protection against bias, misinterpretation, and the compounded cost of a single misstep.

And that cost is not abstract. The numbers do not lie. In 2024, women in the United States earned just 85% of what men earned, a gain of only four percentage points since 2003. Even among 25- to 34-year-olds, the generation most likely to use AI, the gap still sits at five cents.[96]

When decisions about promotion, compensation, and opportunity are shaped by opaque systems and informal judgment, every miscalculation compounds economically. If we don't close this pay gap in the analog world, the AI economy will carry it forward and expand it.

History shows us something important: economic shifts don't automatically level the playing field. In the nineteenth century, insurance companies began replacing intuitive risk assessment with actuarial tables that calculated probability using aggregated data.[97] Once risk became quantified, pricing decisions followed the table rather than the individual story. Judgment moved from discretion to standardized formula. And once pricing followed a formula instead of a personal narrative, inequality became embedded in the system itself.

Economic shifts amplify the structures already in place. The Industrial Revolution made those who already owned factories richer, faster. Financial markets reward those who already hold capital. Technology has never redistributed power on its own; it's scaled whoever was positioned to benefit. If we don't intentionally redesign it, AI will simply widen what's already inequitable.

If economic systems amplify the logic they are built on, then the quality of leadership decisions becomes a structural variable.

Why Leaders Make Suboptimal Decisions

Even the most seasoned leaders aren't immune to making poor decisions. Research and expert observations consistently show that suboptimal choices rarely come from a lack of intelligence or experience. More often they result from predictable patterns in how leaders frame problems, process pressure, and move too quickly to action.

Solving the Wrong Problem (Symptom vs. Cause)

One of the most common breakdowns in decision-making happens before a solution is ever proposed. Leaders rush into execution mode and end up addressing only surface symptoms rather than the underlying problem. Research published in *Harvard Business Review* notes that even highly capable teams "often solve the wrong problems," not because they lack skill but because they fail to pause long enough to define the issue correctly.[98]

"Even the most seasoned leaders aren't immune to making poor decisions. Research and expert observations consistently show that suboptimal choices rarely come from a lack of intelligence or experience."

Wedell-Wedellsborg's research shows that defining the problem is often the hardest part of solving the problem. When leaders move fast and misdiagnose the issue, they can deploy significant resources, talent, and effort toward solutions that never address what's actually driving the challenge. For women leaders, this risk can be amplified. The pressure to demonstrate decisiveness and competence, especially in high-visibility roles, can make it tempting to move quickly toward answers. But speed without clarity increases the likelihood of solving the wrong problem efficiently.

Reframing requires slowing down long enough to ask a deceptively simple question: "*What problem are we really trying to solve here?*" Without that pause, leaders may expend energy on the wrong targets, miss critical context, and compound decision fatigue and especially costly outcomes in roles where scrutiny is high and margins for error are thin.

Narrow Framing and Binary Choices

Another frequent trap is framing decisions too narrowly. For example, seeing choices as yes-or-no binaries when other creative options exist. However, research by Paul Nutt at Ohio State University found that when managers generated at least three credible options, the quality of decisions rose significantly.[99] But in practice, many leaders never get past the binary. Time pressure, decision fatigue, and the expectation to "just make a decision already!" can compress complex problems into false either-or choices. For women leaders, this pressure is often intensified. When decisiveness is closely tied to credibility, slowing down to explore other pathways can feel risky, especially in environments where women already face heightened scrutiny for hesitation or uncertainty.

The danger of narrow framing isn't that you make a bad choice or even that you make the wrong choice. It's that you never see the better ones. When decisions get reduced to a simple yes or no, stay or go, we stop expanding what's possible and, over time, that habit shrinks impact far more than any single wrong call.

Hidden Biases

It's unlikely any of us ever make a truly unbiased decision. We all carry cognitive shortcuts and biases we're often unaware of that quietly shape how we interpret information, assess risk, and choose what to do next. These mental shortcuts exist to help us move quickly, but they can just as easily undermine even the most capable leaders. Bad decisions are made because even the smartest of leaders default to shortcuts, emotions, and blind spots that can distort what options exist.

Overconfidence bias, for example, might lead a leader to overestimate her team's ability to execute on a plan, while confirmation bias can cause anyone to overweigh information that confirms their prior belief and ignore contrary evidence. Studies note that even the most experienced executives fall prey to biases like overconfidence, anchoring on first impressions, and confirmation bias.[100] These hidden biases can skew risk assessments and cause suboptimal choices. There may also be gendered patterns in how bias plays out. For instance, women leaders often have to thread the needle between confidence and perceived "likability," which can sometimes lead to underestimating their own insight (imposter syndrome) or over-indexing on consensus to avoid backlash.

And this is the real challenge with hidden bias: you cannot fix what you cannot see. Awareness helps, but it's not enough. When pressure is high and the stakes are real, leaders need support systems that slow thinking down, widen perspective, and surface blind spots in real time. That is where AI really begins to matter, by creating space for more intentional thinking and expanding the range of possibilities under construction.

Bias becomes more pronounced under fatigue. As the data shows, decision fatigue isn't just exhausting; it's expensive. It costs clarity, erodes confidence, and, in high-stakes moments, it can jeopardize careers.

And for women, the margin of error is expensive. We pay a steeper reputational price when things go wrong. In start-up funding, for example, a 2025 National Bureau of Economic Research (NBER) study shows women are penalized more harshly after failure, are 30% less likely to get funding next time, and raise 53% less capital than male co-founders from the same failed venture. This is clear evidence of harsher judgment for identical outcomes.[101] That's why upgrading the process of decisions, not just the outcomes, is non-negotiable.

How AI Is Empowering Women Leaders' Decision-Making

Leadership decisions don't stand alone. They move through systems, compensation models, performance reviews, hiring pipelines, and capital allocation frameworks, where they become embedded and accumulate over time.

If I go back to the "decision-making era" of my life when I was feeling intense pressure to make the ***right*** decision about staying in corporate or quitting my job, I'm very grateful, because that season of my life taught me what no leadership book ever could: that infinite possibilities exist (even when it feels like there are only two options).

What felt like a personal crossroads was also a structural one. My choice would not only shape my life but also determine how I participated in systems that allocate opportunity.

At the time, I believed my choices were binary: stay or go, push through or walk away, be miserable but rich, or stick to my values but be broke. What I couldn't see because of the pressure, the fatigue, and my own blind spots was that the framing itself wasn't the problem. The options I thought I had weren't the only ones available to me.

That's why I want you to think about AI differently. This isn't a piece of tech that just helps you write a better email. It's a clarity engine that helps you think wider, see straighter, and move forward with greater intention. This is about strengthening your judgement in moments when the cognitive load and hidden biases would otherwise limit your perspective.

Here's how I use AI to do that and how you can too.

Expand Your Horizon

Most of us default to the binary yes-or-no, stay-or-go, speak-up-or-stay-quiet, but life rarely fits into a multiple-choice box. When I was building HER Collective, I thought my options were limited, but by asking, "What are three other unique ways I could approach this?" AI unlocked multiple different pathways for me to explore—ones I didn't even know existed.

What if I tested it as a six-month pilot while staying employed? What if I negotiated a flexible or fractional role instead of a full departure? What if I build strategic partnerships that allowed me to grow the community without carrying all the operational weight myself?

None of those options was visible to me before. I was too close to the problem, but AI helped me widen the frame. It didn't tell me what to choose. It simply revealed that I had more than two doors in front of me.

That shift alone reduced the emotional pressure I felt overwhelmed by. And when you see more options, you make better decisions.

Check Your Bias Before it Checks You!

Our brains are MESSY and filled with unconscious bias that impacts our decision-making. AI can help you catch those blind spots. Ask it to challenge your logic: "Where might I be wrong?" or "What am I not considering?" are great questions to ask to challenge your perspective directly and help you uncover those hidden biases.

Finding the Data that Matters

We're all drowning in information overload. Oftentimes we don't make a decision because we have too much information to sift through. Now, you can upload your information into AI, ask it to analyze, pull out key points, and identify the data that matters, so you can free your brain up to do what it does best—lead like a woman.

Keep your Decisions Alive

What feels like a small weekly ritual is quietly preventing larger misalignment from taking root.

A decision only matters if it stays in motion. Lately, I've been experimenting with using AI as an accountability partner to help me close that gap between intention and execution. At one point, I realized I was trying to do all of my strategic thinking alone while making hundreds of decisions a week (but we all know that isn't sustainable or healthy).

So, I created what I call **HER Board,** a Custom GPT that's a personal board of directors made up of leaders whose judgment I trust and whose standards are higher than my own on the hardest of days. Women and men like Mark Cuban, Natalie Ellis, Jenna Kutcher, Gary Vaynerchuk, and Indra Nooyi. People who think long-term, operate with clarity, and understand scale are some of the people I admire most, and who I believe can help me continue to scale my business.

Of course, we don't actually sit around a table—this is AI, and I'm not that "big time" **YET**—but at the start of each week, I sit down with HER Board and open up all the decisions I have swirling around in my head. This is the brainstorming. I bring the question, the constraint or

"A decision only matters if it stays in motion."

the fork in the road, and then I deliberately expand my thinking with my board by asking: *What options am I not seeing? Where could I be overcomplicating this? What would future me choose here?* And I always have one goal in mind: to have a clear decision or direction forward.

Then the week happens. Life flies by as it always does. But at the end of the week, I come back to the same board to update them on my progress. This Custom GPT is programmed to hold me accountable and ask me questions such as: *Erica, did you act in alignment with what we decided? Where did you follow through? Where did you drift this week? What did you learn that should inform the next decisions?* They are basically asking me: ***Did you do what you said you were going to do, Erica?*** This is where accountability lives within AI. Because this one weekly rhythm I have established—to bring the decisions that need to be made to the table, make those decisions deliberately, and then to come back and close the loop—keeps momentum alive without burning myself out (my board is also programmed to know I tend to inflate what I can do in a day, so they constantly check me).

HER Board keeps me going. It's where decisions are both made and carried out. And the decisions don't stop when the meeting ends. They follow us home.

Decision-making, Motherhood, and AI

This isn't just about strategic decision-making in the boardroom. It's also about the daily decisions and mental load we carry each day. My friend, Tara Ryan, is the co-founder of Confidante, with Ariel Galipeau. Confidante is an AI tool that honors your way of mothering when the mental load feels like too much. It offers a way to offload some of the hidden decisions—like "What do I tell my boss?" "When do I say no?" and "How do I set boundaries?"—and gives women a tool to turn mental churn into clarity.

According to Tara Ryan, it's all about "values-based decision-making." Values-based decision-making is simply the practice of aligning your choices with your core beliefs, priorities, and personal definition of success, not just external pressure or short-term payoff. So, when a mother is faced with a decision: do I go on the business trip and miss my child's performance at school, or do I stay home and miss the opportunity, Confidante will be there to guide her and help her carry the load.

The Real Power of Decision

As a woman leader, your decision doesn't just impact you; it ripples across your team and your family. Having a tool that can hold the context, track what matters to you, and prompt you with questions that you didn't think to ask gives you the clarity needed to make the best decision possible—a decision grounded in alignment.

Because at the end of the day, every great leader is defined by one main thing: the quality of their decisions. And in this new era, the leaders who will rise to the top are the ones who will know how to combine data with discernment. They will be the leaders who use AI as a strategic advantage and a decision-making tool.

When we lean into AI as our decision partner, we stop reacting impulsively and start architecting our future. We make choices that are rooted in clarity instead of a hunch. And we can start to build careers and lives that are aligned with our deepest values instead of someone else's definition of success.

Having a seat at the table taught me something I didn't expect. The seat alone doesn't change anything. The direction of the room changes when the decisions do.

Every decision you make shapes more than the moment. In organizations shaped by data and automation, leadership choices become part of the structure. They influence how performance is measured, how compensation is allocated, and how opportunity is distributed.

AI carries those decisions forward. It turns judgment into patterns. Over time, those patterns become standards. And standards become infrastructure. Infrastructure does not forget. Once a pattern is encoded, it travels through systems, into dashboards, into reviews, into compensation models, into the next generation of leaders. The way you decide today becomes part of the structure others inherit.

TLDR

- The quality of your decisions determines the quality of your leadership. Most leaders don't fail from a lack of vision, but they fail from decision fatigue, bias, and poor execution.

- Women are judged more harshly for missteps, which means we're not just making decisions under pressure; we're making them under scrutiny that compounds every outcome.

- AI is a tool that expands your options, checks your biases, and strengthens your follow-through.

- When you combine data with discernment and let your values lead, you make decisions that move both your strategy and yourself forward.

YOUR AI-POWERED ACTIONS

ASK:

What decision in your life or leadership feels heavy right now, and how might AI help you see it with fresh eyes?

APPLY:

Use AI to map three (new) possible paths for an upcoming decision. Ask it to analyze each path for risk, opportunity, and alignment with your top three core values. Then check in with your intuition. Which path feels right?

AMPLIFY:

Share your process. Talk with your team or your family about how you used AI to make a stronger values-based choice. Then, create a chat agent trained on the top decision-making matrices that exist to help you make future decisions.

CHAPTER 6

MOMENTUM IN MOTION

October 1, 2024, was my official last day in corporate America. I wrapped up my transition documents, said my final goodbyes to my team, and logged off forever. The next day, after I dropped the kids off at school, I sat down at my desk to start my full-time role as a CEO and founder, as an entrepreneur, and I truly had no idea what to expect.

And in that moment, as I opened my laptop, the quiet was deafening. There was no urgent employee issue waiting for me. There were no mandatory training sessions from our parent company or last-minute "urgent" requests. There were no meetings on the calendar and no built-in deadlines. I had my ambition, my mission, and no clue what would come next. I also had two kids (ages nine and six at the time) and a brain that loves to spiral. In that first season, I learned a lesson that I've carried with me ever since: momentum is not about doing more. Momentum is about installing rhythms that keep what matters moving when life gets wild. The leaders who build rhythm now will not be scrambling later. Systems reward those who practice before the rules fully change.

I have to admit, in the beginning, I was a bit scattered. But it was okay because I was in build mode. Yet, when the noise of kids, clients, launches, and life started to creep back in, I went looking for support that could hold me to a rhythm. That's where AI comes in. As of the writing of this book (November 2025), a feature of ChatGPT Pro is what they call a "daily pulse." (Pulse is a feature in ChatGPT that provides a brief summary of your recent activity and current context. It helps users quickly understand what they have been working on so they can resume tasks or make informed decisions about what to do next.) I have curated this daily pulse to set me up for success from the jump, and it's the first thing I look at before I start my day.

The work that produces meaningful results is typically the result of consistent effort. You know this from fitness. You don't see results after one workout, but from showing up again and again, even when you don't feel like hitting the gym. Over time, that consistency starts to show.

This specially curated pulse helps me stay focused on my goals, reduce mental friction, track honest progress, and anchor myself in abundance and possibility with daily meditations and exercises that ground my nervous system. Consistency is what makes the signal visible. And I'm still tweaking my practices. The tech will evolve, maybe before you finish reading this book. And so will we. The point isn't perfection, but a repeatable rhythm that feels sustainable and human.

Why Rhythms Help

Rhythms turn good intentions into repeatable motion. When Henry Ford introduced the assembly line, productivity didn't increase because workers were putting in more effort. It increased because repetition was systematized. Tasks were sequenced. Cues were fixed. Output became predictable.[102] Rhythm, once embedded into the process, scaled production. Leadership momentum follows the same principle. Once action becomes structured, output starts to build. The energy cost of consistency drops because small actions are anchored to real-life cues. Instead of constantly re-deciding, there is a clear next step. And the mental load shifts so your attention stays on the work that matters.

Most important rhythms honor recovery as part of performance, not a reward for it. AI strengthens the whole system by capturing what you offload, nudging you at the right moment, and reflecting back your wins so you can adjust without judgment.

The Momentum Rhythms

Core rhythms create sustainable momentum for women leaders in the era of AI. I've identified three focused rhythms to start with for maximum benefit, and once you integrate these into your life, you can add more if your season calls for it.

- The Rhythm of Focus
- The Rhythm of Support
- The Rhythm of Recovery

Rhythm 1: The Rhythm of Focus

The rhythm of focus is all about choosing what matters most and building momentum one move at a time. Then repeat. Here's why it matters.

Every success story, whether in leadership, health, or business, comes down to one thing: the actions you take consistently over time. Big goals are built on small, steady moves that become habits, and those habits ultimately determine your results. It isn't luck or timing that creates success; it's rhythm.

But sustaining those rhythms is hard because motivation fades and life gets loud. There's something else worth naming here: access to

---

"The tech will evolve, maybe before you finish reading this book. And so will we. The point isn't perfection, but a repeatable rhythm that feels sustainable and human."

tools doesn't automatically translate to advantage. New tools will not change your trajectory if you hesitate to use them. Harvard researchers found that women adopt generative AI at rates roughly 25% lower than men.[103] In many organizations, experimentation is framed as innovation for some and risk for others. When cultures penalize women more harshly for visible mistakes, hesitation becomes understandable, and it also becomes costly.

That gap matters. Underrepresentation in AI usage and in AI training datasets means systems are built on narrower inputs and narrower experiences. The people who experiment early help shape the patterns that scale later. Momentum, then, isn't reckless risk; it's deliberate exposure. It's practicing now so you aren't sidelined when the systems mature.

And this is exactly why rhythm matters! Habits matter. Automate what's important so you can keep moving forward even when your energy dips. Despite what productivity culture sells, there's no magic day count for habit formation.[104] The strongest evidence shows that habit strength consolidates over months, not days, and varies by person and behavior. What sticks is not intensity but repeatable action anchored to consistent cues. Early repetitions deliver the biggest gains. And self-chosen behaviors practiced at predictable times strengthen faster.

How AI Supports Your Focus Rhythm

I think we all know how frustrating it can feel when you aren't focused but you know that your focus is required. You sit down to work, but your attention is fractured. You're pulled by notifications, unfinished decisions, competing priorities, and the quiet anxiety that you're forgetting something important. You're busy doing a lot of things, but you're not really getting anything done. Without focus, forward momentum becomes almost impossible.

But here's my unique twist. I don't believe that focus is a discipline problem or a personality trait. It's a systems issue.

When your attention isn't protected, even the best decisions stall. When you're dialed in, progress compounds.

This is where AI becomes genuinely supportive because it can help you design a rhythm that keeps your attention intact, one that protects your priorities before the week begins, supports you while you're doing the work, and helps you close decisions instead of carrying them into the next week. That rhythm has three parts:

- The tools you choose
- How you open the week
- How you check in and how you close the loop

Build a Focus Stack that Protects What Matters

Your Focus Stack is a set of AI tools or software that protects your attention from planning through execution, so that your best work actually gets done. Think of it less like a tech stack and more like a **decision support system for your time.**

Scheduling tools defend your priorities by blocking focus time, rescheduling intelligently when the day shifts, and keeping high-value work from getting crowded out. **Planning tools** help you decide where your energy goes each day. **AI note-taking tools** turn conversations into clear next steps instead of mental clutter. **Workspace assistants** bring everything together, so clarity lives in one place.

Each tool has one job: reduce friction, surface priorities, and keep you aligned with what matters most. You don't need **EVERY AI tool out there.** (Please, don't go out and spend thousands of dollars a month on all these tools.) Just pick one or two that make your life easier. Start there and keep only what works.

BUILDING YOUR FOCUS STACK WITH AI TOOLS

Build a Focus Stack that protects what matters and reduces friction. Start with one or two and keep only what helps you.

1. Scheduling with AI that defends priorities and protects boundaries:
    - Calendar assistants autoblock focus time for your top work, reschedule intelligently when the day shifts, and keep at-risk tasks visible so the important work always has a slot on the calendar. Popular AI scheduling tools are Calendly AI, x.ai, Google Calendar + AI Enhancements, and Motion.
    - Planning tools can guide a calm daily review and time-box tasks, so you decide what gets your energy today. When it comes to planning tools, there's no shortage of options to choose from, including Notion AI, ClickUp AI, Asana with AI, Miro, and Trello.
2. AI note-taking that plans your next move:
    - AI notetakers can turn meetings into summaries with action items assigned to each person that you can drop straight into your planner. There are no more sticky notes or "I'll remember this later" math that needs to be done. Popular options as of the writing of this book are Otter.ai, Granola, Fathom, and Fireflies.ai, but many meeting platforms, such as Zoom, have notetakers built in. Granola is my favorite personal note-taker.
3. Clarity all on one screen:
    - Workspace assistants like Notion AI, Google's or Microsoft's built-ins condense long notes into the day's top three, and generate a 10-minute first step for each.

Open the Week with Clarity

At the start of each week, I use AI to lay out what decisions need to be made. This is where I clarify what matters most, pressure-test priorities, and decide what success looks like when I follow through. Mondays for me are all about asking:

- What actually deserves my attention this week?
- Where am I at risk of defaulting to urgency?
- And what would alignment look like if I protected it on my calendar?

Midweek Check-in: Stay Aligned While the Week Is Flying By

By Wednesday, most plans are under pressure, and I don't know about you, but I often feel behind and worry I won't hit my targets. This is where a short, midweek check-in keeps me focused. I use AI to quickly assess what's moving, what's stuck, and where reality has shifted since Monday. It's very plausible that on Tuesday, I got a call from the school to pick up a sick kid—and now my week is trash. My time here is spent recalibrating by asking:

- What still matters?
- What needs to be adjusted?
- What can I release without guilt? (This one is the hardest!)

This midweek check-in helps me spot misalignment early, while I still have time to correct the course. My focus stays flexible, and I stay on track.

Close the Loop and Clear the Load Fridays

A decision only matters if you follow through on it. At the end of the week, I use AI to review what actually happened. This is where I name my wins, capture loose ends (that I might forget about over the weekend), and decide what to do with anything unfinished. Some decisions get recommitted, while others are consciously moved down the priority ladder. Nothing stays vague or mentally "open" in my head.

I want to pause here for a moment of coaching. If you've ever ended a long week feeling completely depleted, swamped, stretched thin, and somehow unsure what you actually accomplished, listen closely. This is the mindset shift that changed everything, and AI can help you practice it.

Fridays are for release. Use AI to review what moved forward, what didn't, and why, and what should inform your next decision. Set down the guilt. Pick up the belief that closing the loop isn't failure or pressure, but it's true leadership. When decisions are named and completed, your brain no longer has to carry unfinished business into the weekend or the week ahead. Reset your focus and let AI hold the details. Keep your momentum moving forward.

"When decisions are named and completed, your brain no longer has to carry unfinished business into the weekend or the week ahead. Reset your focus and let AI hold the details. Keep your momentum moving forward."

THE FOCUS RHYTHM SYSTEM

The Focus Rhythm System is a practical cadence that turns intention into execution without burning you out. It prepares you to start the week with clarity, checks momentum before it slips, and closes the loop so unfinished work doesn't follow you into the weekend.

Use this rhythm during seasons when you're leading, building, or carrying a heavy mental load. When your calendar is full but progress feels unclear, this cadence helps you convert intention into visible movement without sacrificing your energy or your boundaries.

Instead of trying to manage everything, you'll practice naming what matters, anchoring action to your real calendar, and making clean decisions to continue, defer, or let go. Over time, that consistency builds self-trust. You begin to believe in yourself because you keep showing up in ways that are sustainable and actually fit your life.

The Monday Alignment Brief

ROLE: You're my strategic thought partner and operations-savvy coach.
GOAL: Help me open the week with clarity by identifying the few priorities that truly matter, defining what "done" looks like for each, and anchoring the first step to my real calendar, so progress actually starts.
CONTEXT: I will paste my calendar, deadlines, and rough notes below. I protect early mornings for deep work, school pickup is fixed, and I want visible progress on thought leadership and revenue activities.
CONSTRAINTS: 150 words max. Plain language. Anchor every first step to an existing meeting or time cue. Include one item to drop or defer. Suggest calendar holds I can paste in. For each priority, define the outcome, a 10-minute first step, and a specific calendar cue. Name one item to defer or drop. Flag one boundary I need to protect this week.

CONFIRMATION: Ask up to five questions, one at a time, to confirm what's truly mission-critical, any non-negotiables, and where a boundary is needed before drafting.

The Midweek Momentum Check

ROLE: You're my grounded accountability partner and momentum-focused coach.
GOAL: Help me assess progress midweek, surface any risks early, and make clear practical adjustments so priorities stay alive and momentum continues.
CONTEXT: I will paste my done list, new requests, and any changes to my calendar since Monday. I'm balancing delivery, decision-making, and long-term priorities, and I want an honest read on what's moving and what's at risk.
CONSTRAINTS: One page. Plain language. Summarize momentum in 100 words or less. Identify one priority at risk and ask why. Recommend two 10-minute actions I can take today.
CONFIRMATION: Ask up to three questions, one at a time, to confirm priorities, capacity, and where a boundary or renegotiation is needed before recommending next moves.

The Friday Release

ROLE: You're my grounded accountability partner and operations-savvy coach.
GOAL: Help me close the week cleanly by naming progress, deciding what's finished, and making clear choices about anything unfinished so I can release the mental load and start the next week focused.
CONTEXT: I will paste my calendar, done list, and any open threads from this week. I've been balancing leadership responsibilities, revenue-driving

work, and long-term priorities. I want clarity, not judgment, about what moved and what didn't.

CONSTRAINTS: One page. Plain language. Summarize the week in less than 100 words. Name three wins with receipts. Identify what's complete, what remains open, and why. For anything unfinished, choose: recommit (with a 10-minute first step and calendar cue next week), defer, or drop. End with one boundary or rest practice to protect over the weekend.

CONFIRMATION: Ask up to four questions, one at a time, to clarify what counts as a win, which receipts matter, and answer any questions you may have.

Rhythm 2: The Rhythm of Support

The rhythm of support is all about offloading the mental load and replacing willpower with structure and support. Why is this important? Women leaders carry an immense amount of invisible work. Cognitive unloading, which is the moving of tasks, reminders, and context out of your head and into reliable systems, is what reduces the mental strain and prevents the start/stop cycle. The cognitive science is clear. If we offload the mental burden to external tools or prompts, we cut the processing demand and perform better on what matters most.[105]

How AI Supports your Support Rhythm

Support is not a luxury. As a leader, especially a woman leader, the mental load is heavy. Use AI as the structure to keep your best work in motion.

Capture

Voice-dump ideas, follow-ups, and loose ends into AI after meetings. Ask it to tag by project and person. Use this feature when you're commuting or doing other random tasks to prevent forgetting the important things!

Clarify

Have AI sort your capture into these four categories, which I call the 4P Method: **Perform** (something you and only you can do), **Pause** (put this on the back-burner), **Pass** (delegate this to someone else), or **Purge** (this is not necessary and does not align with the goals you have outlined). Ask it to draft the first pass of messages or checklists so you can move your work forward.

PERFECT PROMPT FOR YOUR SUPPORT RHYTHM

Use this prompt when your notes are messy, meetings have multiplied, and you know what matters but can't see a clear path to execution. It's perfect for high-stakes, multi-owner weeks when you need to reduce the cognitive load, create clear ownership, and move the work forward without overthinking or over-explaining.

ROLE: You're the chief of staff for my brain and a calm, systems-minded project manager.
GOAL: Convert the notes I paste below into a clean, execution-ready support plan that reduces cognitive load and keeps multi-owner work moving. Organize actions by project into a two-column format, naming the owner, true due date, effort estimate (10, 25, or 60 minutes), dependency, and stoplight status for items owned by me. Generate two 10-minute micro moves. Recommend the next realistic calendar slot and draft a first pass message I can paste into email or Slack to advance the work without over-explaining.
CONTEXT: I lead cross-functional work with internal partners and contractors. I already know what matters. I need help translating shared notes into coordinated follow-through. I work best when momentum starts small. Mornings are protected for deep work. Afternoons are meeting-dense, and my systems of record are Notion and Google Drive.
CONSTRAINTS: My actions must start at 10 minutes. Tag each item as Do, Delegate, Decide, or Defer. Keep the tone clear and human. Output must be paste-ready today. Flag at-risk items and propose one delegate or re-scope option if the scope is unrealistic. Schedule a Wednesday afternoon reminder for the top three actions and include a two-line Friday recap message I can send to stakeholders.
CONFIRMATION: Ask up to five questions, one at a time, to confirm real owners, non-negotiable deadlines, priority order, critical dependencies, and what "done" actually means before producing the plan.

Commit

Have AI convert tasks into cue-anchored micro-moves: small actions tied to something that's already on your calendar, so execution doesn't rely on memory or willpower. When actions are attached to real moments, your brain can stop holding them. For example, "After the Status Update Meeting, I'll send two sponsor updates." Commit reduces the start/stop cycle by letting structure do the remembering.

Check-In

Schedule a weekly 10-minute AI Review. Ask: what moved? What stalled? What's the one bottleneck to remove? This check-in provides cognitive support intended to keep you focused. AI reflects patterns you might normalize, surfaces constraints you're adapting to without noticing, and helps you decide what deserves your attention next. Support in this context means continuity. The work stays visible, decisions stay alive, and your energy stays focused on moving forward.

Rhythm 3: The Rhythm of Recovery

Burnout is recognized by the World Health Organization as an occupational phenomenon tied to chronic workplace stress, and what I've learned the hard way is that the recovery I once viewed as an indulgence is really a strategy to keep you at your best.[106] The ability to psychologically detach from work during non-working hours is associated with better mood, energy, sleep, and overall well-being—*go figure!*[107] So the rhythm of recovery is a matter of protecting the engine. Why? Because recovery sustains momentum.

How AI Supports Your Recovery Rhythm

Energy Diary

Most people think that recovery starts with rest. But I believe that recovery starts with awareness. Most burnout doesn't come from doing too much in one day. It comes from repeating that same energy-draining pattern day after day, week after week, month over month, quarter over quarter. Without insight into what drains you and what restores your energy, recovery becomes reactive instead of intentional. An Energy Diary turns recovery into a light, repeatable practice. In just a few minutes at the end of the day, AI helps you capture how your workday actually impacted you. What gave you energy, what cost you energy, and what do you want to adjust next?

Ask AI for a five-minute end-of-day reflection with three questions:

- What energized me?
- What drained me?
- What will I do differently tomorrow?

Over time, the small check-ins compound, and patterns start to emerge. You start designing weeks that protect your energy instead of depleting it, and that, my friend, is when momentum becomes sustainable. Each month, you can ask AI to pull your patterns and analyze your energy spikes and dips. This insight becomes your playbook for planning weeks that support your life and the goals you want to achieve.

Boundary Creation

When I talk about boundaries, I'm talking about clear expectations that protect your focus, your energy, and your priorities without apology or over-explanation. When you put a boundary in place, it works like a porch light, showing people where the door is and how to approach you. It creates clarity around your capacity, timing, scope, and availability so that your commitments align with what truly matters. Many leaders struggle with boundaries, often because they know what they need, but they don't know how to say it in a way that feels professional, kind, and firm at the same time. This is where AI becomes supportive. It helps you translate intent into language so that you're not negotiating real time or defaulting to a "yes" when you really want to say "no."

When you need to set a boundary, have AI generate one boundary script you can use as a template with colleagues or clients. These scripts remove emotional labor from the moment and give you a steady, repeatable way to protect your time. For example, if your colleague asks you to review their slide deck but you're strapped for time, a simple boundary script might sound like, "I'm heads down on priority work this morning, but I can review this after 2 p.m. or tomorrow. Let me know what timing works best." A simple professional boundary template that you can use in almost any situation sounds like this:

"I'm at capacity this week, and I want to give this the attention it deserves. I can revisit this on [date], or we can identify another owner. Let me know what works best."

My personal favorite is a creative out-of-office reply that people actually read: "I'm currently on the beach with a mocktail in hand, no laptop in sight, and my nervous system unclenched. I'll be back on {date} and will respond then." (Yes, I've really used this in real life and I always get the best responses!)

Rest Rituals

Build a tiny shutdown ritual that closes the loop on your day instead of carrying it into the night with you. Use AI as a gentle container for the

"A simple professional boundary template that you can use in almost any situation sounds like this: 'I'm at capacity this week, and I want to give this the attention it deserves. I can revisit this on [date], or we can identify another owner. Let me know what works best.'"

PERFECT PROMPT FOR YOUR RECOVERY RHYTHM

Rest and recovery are responsibilities you carry as a woman leader so you can keep showing up with clarity, courage, and conviction. Your future output is built on today's recovery. Here's a prompt to support you.

Use this during seasons when you're carrying a lot, building something meaningful, or feeling the early edges of depletion. This rhythm keeps you aware before you're overwhelmed.

ROLE: You're my burnout-prevention coach and reflection partner. You blend emotional intelligence with practical structure. Your job is to help me slow down long enough to notice what's working, what's costing me energy, and what needs to shift so I can lead sustainably.
GOAL: Create a simple, five-minute end-of-day reflection I can answer by voice. Your questions should help me name my wins, my challenges, my energy trends, and any adjustments I want to carry into tomorrow. At the end of each week, you will compile these answers into a clear summary of patterns, insights, and recommendations.
CONTEXT: I'm a high-achieving entrepreneur with a fast brain, a full life, and a tendency to power through until I hit my edges. I want a rhythm that keeps me honest, grounded, and aware of my capacity—without adding more work to my plate. This reflection needs to feel light, repeatable, and supportive, not heavy or dramatic.
CONSTRAINTS: Questions must be simple, consistent, and answerable in under five minutes. Prompts must work seamlessly with voice input—no long explanations or complex branching. Each question should map cleanly to a weekly summary template you can produce every Friday. Keep the tone warm, concise, and human. No more than five questions per day.
CONFIRMATION: Before creating the final reflection, ask me five to seven questions to personalize my experience.

process so you're not deciding what to reflect on or what comes next. You're simply noticing and then letting go. Ask AI to prompt a brief end-of-day reflection by naming:

- Three wins you want to acknowledge (even the micro ones!)
- One gratitude that softened or supported you today.

Then, hand off the mental load. Have AI identify your true top priority for tomorrow and schedule it for you, so your mind doesn't have to keep rehearsing what's next once the workday is over.

When You're Ready for More: Rhythms That Deepen Your Momentum

One thing I've learned in the last decade of my life is that some seasons demand simplicity, while others invite expansion. I felt compelled to include a prompt on rest and recovery because, for the bulk of women I coach, rest is always a challenge. If you're in a moment where you have the capacity for more support, more structure, or more intention, layer in one rhythm at a time. (And we always have capacity for more support!) Do this slowly and consciously. These additional rhythms are not simply "add-ons" but enhancers and amplifiers. They are the quiet systems that can keep it all together for you while you grow.

Rhythm of Connection

Every leader has a personal board of directors: the inner circle who lifts, celebrates, and studies you. But when life gets loud, staying connected takes intention. Track your top 10–15 relationships and let AI help you keep those connections warm. Use AI to draft a thoughtful check-in, remind you of the last conversation, and even prepare one meaningful question you can ask the other person.

Rhythm of Visibility

Momentum isn't just about doing the work. It's about being seen doing the work. Once a month, batch your ideas and pick one strong insight to ask an AI to shape it into a LinkedIn post, a tight two-slide carousel, and a 150-word newsletter. Review it to ensure your voice stays authentic, but let AI build the scaffolding so you can show up consistently without burning out.

Rhythm of Creativity

Give yourself 45 minutes each month to get creative and play around! Follow where your curiosity leads you. Use AI to sketch outlines, explore different angles, or draft the very messy first version of a new offer or presentation. Creativity is often the oxygen that fuels our excitement and our passion. Give yourself a dedicated time block each month to make that exploration possible.

Rhythm of Learning

Growth often feels like a full-time job. But with the power of AI, you can create quarterly micro-learnings tied to one skill that will actually move your career forward. You can have AI pull credible research, build practice exercises, and keep your development intentional, but manageable.

What Momentum Feels Like

You will know your rhythms are working long before the results show up on paper. You actually feel it in your body first. Momentum is when getting started feels easier than skipping. When the things you used to avoid become the things you knock out without the added procrastination. When your to-do list becomes the "look what I've already done" list.

Momentum is that moment you realize you no longer have to push as hard as you used to because the system you have created is carrying some of the weight. And that's the moment you know that you aren't just DOING the rhythms but are being shaped and changed by them. And that's the whole point of this chapter.

You don't need to work harder or put in more effort. You need rhythms that make your effort sustainable. AI doesn't create your momentum. It carries forward what you repeat, one micro-move, one cycle, one season at a time.

In environments shaped by automation and predictive systems, repeated actions turn into signals that drive visibility and shape momentum over time. That momentum doesn't just stay internal; it leaves a trace that the system can recognize and respond to.

TLDR

- Momentum isn't intensity; it's rhythm. When your life gets loud, rhythms give you a way to keep the important things moving forward.

- AI strengthens your rhythms by reducing friction, capturing details, and surfacing what matters.

- Start with three foundational rhythms: focus, support, and recovery.

Add optional rhythms only when you have the capacity: connection, visibility, creativity, and learning.

- Momentum becomes real when getting started is easier than skipping.

- Sustainable leadership comes from systems that support you, not from doing more or pushing harder.

YOUR AI-POWERED ACTIONS

ASK:

Where in your life or leadership do you feel the most friction right now? Is it clarity, follow-through, energy, boundaries, or simply the weight of carrying it all?

APPLY:

Choose one rhythm (focus, support, or recovery) and install the simplest version this week. Anchor it to a real cue, keep it small, and let AI take the mental load wherever possible. Give it seven days. Let it build its own momentum.

AMPLIFY:

Once that rhythm feels natural, layer in another. Use your new capacity to stretch your visibility, creativity, or connection.

PART 3

CLOSING THE COMPOUNDING GAP

Turning Income into Leverage

Authority changes how you are seen. Leverage changes what becomes possible.

Compensation reflects how your work is seen, valued, and carried through the system. It shows how organizations price scope and how influence turns into opportunity. And when that value builds unevenly, you see it in who gets opportunities, who has options, and who has the freedom to decide what comes next.

The Compounding Gap is the distance between contribution and ownership. It shows up in pay disparities, in negotiation outcomes, and in who accumulates assets versus who remains dependent on a paycheck. Over time, it widens quietly over time through missed raises, undervalued scope, interrupted career paths, and financial decisions made without clear data.

Closing this gap requires more than confidence. It requires clarity about market value, documentation of measurable impact, and strategic negotiation grounded in evidence. It requires understanding that income is participation, while wealth is infrastructure. Income moves month to month. Wealth compounds across years. One funds your life. The other funds your options.

This section shifts the conversation from earning to engineering. You will examine how compensation systems function, how bias can surface inside negotiation, and how to use data and AI to make your value visible and defensible. You will design financial systems that automate stability and build optionality long before it feels urgent.

Wealth does not accumulate through intention alone. It grows through structure. Through consistent decisions that redirect money toward assets, protection, and long-term leverage. When those systems run quietly in the background, freedom becomes less emotional and more architectural.

The Compounding Gap narrows when contribution turns into ownership. When income becomes investment. When financial decisions align with the life you're building rather than the one you're trying to escape.

This is where effort turns into equity. This is where leverage begins to compound.

CHAPTER 7

THE VALUE EQUATION

When I look back on my corporate career and every salary negotiation I ever walked into, the early ones were fueled by one thing: desperation. I didn't negotiate because I understood my value; I negotiated because someone once told me, "Never take the first offer. Always ask for more." So, if they came in at $60,000, I'd muster the courage to ask for $65,000. Not because I believed I was worth it, but because I didn't want to leave money on the table.

But as my experience grew, and as I learned what real negotiation actually requires, everything shifted. I stopped negotiating from a place of fear. I stopped guessing, and I stopped treating salary conversations like a gamble. I started negotiation from data, evidence, and proof of impact. That's what changed everything. The difference wasn't confidence; it was evidence. And once I understood that compensation follows what can be measured, everything changed.

What I thought was a personal confidence problem was actually a structural one. Compensation systems are measurement systems, and what gets measured is what gets rewarded.

In this chapter, we will dismantle the myths that keep women underpricing, name the bias that shows up in pay conversations, and make AI your ally that can pull hard numbers for you and draft language you can practice until it sounds like you.

Negotiations take confidence. Confidence requires clarity. Clarity sounds like numbers.

And the numbers prove this is not anecdotal.

The Reality Check

Let's start with the absurd reality that women in the United States still earn less money than men do. Using weekly wages, women earned about 83.2% of what men earned in the fourth quarter of 2024. Looking at hourly earnings across full and part-time workers, women earned about 85% of men's pay in 2024.[108] Mothers face an even steeper penalty, earning roughly 74 cents on the dollar compared with fathers when both work full-time, year-round.[109]

There's debate about the "ask gap." Some newer evidence shows women, especially among MBAs and in specific contexts, are negotiating as often or more than men, yet we still see lower outcomes.[110] Other research finds men are more likely to negotiate first offers, particularly right out of college, and that information gaps about negotiation gains keep many women on the sidelines.[111] The takeaway is simple: this isn't about confidence, but about data, design, and bias. And when design determines outcomes, negotiation stops being personal and becomes structural.

Bias still shows up at the table. A large body of research led by Hannah Riley Bowles and colleagues finds that when ambiguity is high and standards are vague, women who negotiate can face social penalties.[112] If we can reduce the ambiguity, we can reduce the bias. When we bring objective standards and a clear scope, the playing field becomes fair.

And we cannot ignore the broader structural reality: representation at the top remains deplorable. Women make up roughly 22–30% of the AI-related workforce and hold only about 12% of AI research positions.[113] When the systems shaping compensation, promotion, and technological infrastructure are still overwhelmingly designed and led by men, neutrality is an illusion. When system designers remain homogeneous, the systems they build reflect those blind spots in what gets measured and rewarded—something we can't allow to go any further.

As you negotiate your value, remember this: you're not entering a system that was built with you at the center. You're entering one that was historically calibrated around someone else. That isn't a reason to shrink. It's a reason to arrive prepared. Preparation is how individuals operate strategically inside systems they didn't design. Use AI tools to amplify your achievements, sharpen your evidence, and build leverage that challenges old hierarchies instead of adapting quietly to them.

You know what helps? Pay transparency and data.

Laws requiring salary ranges in job postings are expanding across the United States, and organizations increasingly use pay-equity tools to find and fix the gaps.[114] Transparency alone, however, will not solve everything, and the evidence points in the same direction. When we have more information, we have fairer outcomes, and employers are finally starting to figure this out. When compensation is visible, systems are forced to be accountable.

Now that we know the landscape, we can build out the AI-assisted playbook.

The Value Equation

Here's the hard truth: you don't get paid what you're "worth." You get paid what you can show, scope, and negotiate. Compensation isn't a reflection of effort, and the system rewards what it can measure. If women's impact is undermeasured, it will be underpaid. And as compensation systems become increasingly digitalized, what gets measured doesn't just influence pay; it shapes future opportunity.

In the early twentieth century, Frederick Winslow Taylor was the American engineer who changed the way work was organized. He introduced "scientific management," breaking factory labor into timed,

---

"When the systems shaping compensation, promotion, and technological infrastructure are still overwhelmingly designed and led by men, neutrality is an illusion."

measurable units and tying wages directly to output.[115] Once work was reduced to quantifiable tasks, compensation followed the metric, not the individual narrative. Productivity was defined by the system, and pay was scaled accordingly. Modern compensation systems function the same way. Once contribution becomes data, compensation follows structure. If compensation follows measurement, then we must redesign what gets measured.

This is the model I use with clients and inside HER Collective:

Value = Market x Scope x Outcomes x Scarcity x Risk x Leverage

- **Market** is the current range for your role and location.
- **Scope** is the size, complexity, and span of control.
- **Outcomes** are your receipts: revenue, savings, risk reduced, people developed.
- **Scarcity** is how few people can credibly do what you do.
- **Risk** is accountability you absorb, decisions you own, and the stress you manage.
- **Leverage** is timing, alternatives, and how effectively you frame the ask.

AI supports each factor. It doesn't do the negotiation for you (but by now, you know you can use it to practice!). AI does, however, give you the clarity to negotiate effectively.

Market: Know the Range Before You Walk In

Always start with official data you can gather. Pull the Bureau of Labor Statistics' most recent Occupational Employment and Wage Statistics for your role and your metro. This will provide you with national, state, and metro medians.

Next, look at real-world employer signals:

WHY YOU MUST STILL CHECK THE SOURCES

AI is arguably the most powerful emerging tech today, but it isn't authoritative on its own. Like humans, it can sometimes get things wrong. In AI, these confident but incorrect moments are often called *hallucinations*. **Here's the surprising part: a hallucination isn't a bug**. Hallucinations are a feature and a part of how the system works. Humans do the same thing. When we don't have enough information, our brains fill the gaps with whatever seems plausible, and we're usually very confident about it. AI works the same way. When context is thin or signals conflict, it keeps going anyway. That's why the smartest leaders don't treat AI as the "be-all end-all," but they use it as a fast-thinking partner with a human in the loop.

PROMPT: MARKET BENCHMARKER

When to Use this Prompt: Use this prompt anytime you need real numbers before applying, negotiating, or second-guessing what you're worth. It's designed for the moment you want to replace uncertainty with market-backed confidence and walk into the conversation grounded, informed, and unshakable.

ROLE: You are a sharp-eyed compensation analyst and a labor market insider working for me.
GOAL: Build a one-page market brief that tells me what my role is actually worth right now. Show low, median, and high pay for my target role in my metro plus three recent job postings with posted ranges, so I can see how companies are signaling value in the world. The outcome is confidence before I negotiate, apply, or counter.
CONTEXT: I'm exploring roles and need real market intelligence, not vibes. I'll paste my working title options, location, and three job descriptions. I care about accuracy, clarity, and being able to explain the numbers out loud to a recruiter or executive without sounding unsure.
CONSTRAINTS: Use official sources first, and list each source so that I can find the original citation. Keep to 200 words and one clean table. No filler. No hype. Just signal and insight.
CONFIRMATION: Ask up to three questions, one at a time, to confirm role family, level, and hybrid/remote status before drafting.

- Review salary ranges listed on current job postings in your state, especially if you live in a state with pay transparency laws where ranges are now required and legally vetted.
- Triangulate with market intelligence platforms that combine employee-reported data with employer-submitted or validated compensation information.
- Tools like PayScale and PayFactors sit in this middle ground. They blend crowd-sourced inputs with employer-verified pay structures to reflect how compensation is actually being set.
- Finally, round out your view with broader benchmarks such as Glassdoor's Salary Explorer, Salary.com, and macro-level trackers from the World Economic Forum and OECD that contextualize pay gaps by gender and region.

The goal is to come up with a defensible range that you can speak to with confidence, backed by data employers recognize and respect.

AI can do the heavy lifting of compiling and organizing this information for you, as long as the inputs are credible. Start with verified sources you trust, and then ask AI to synthesize them into a clean, decision-ready format. For example, ask AI to produce a one-page brief that includes your role, location, low, median, and high compensation ranges, and three recent public job postings with listed ranges, all pulled directly from authoritative sources. When you walk into a room with this data, you remove the ambiguity and emotion, and signal that you did your homework.

Scope: Price the Actual Job, Not the Title

Two "Director" roles at two different companies can be two totally different worlds. A Human Resources Business Partner at a Fortune 500 operates quite differently from one at a small, 50-person agency. When evaluating roles, use AI to help you translate the job scope into

PROMPT: SCOPE TRANSLATOR

When to Use this Prompt: Use this prompt when a role sounds senior, but you're not sure the compensation matches the responsibility, or when you want proof before you negotiate. It's especially powerful when evaluating stretch roles, lateral moves, or offers that rely on impressive titles but vague scope.

ROLE: You're a global talent architect and compensation strategist who understands how real authority, complexity, and risk drive pay, not job titles.
GOAL: Translate the job description below into a clear scope profile and an appropriate pay tier relative to my market brief. Surface what the role truly owns, where decisions actually live, and how accountability, scale, and risk should show up in compensation. The outcome is clarity about whether this role is properly leveled or quietly underpriced.
CONTEXT: I will paste a job description and my market brief. I want to see past inflated titles in vague language to understand what the company is really buying and what I'd be responsible for delivering.
CONSTRAINTS: 150 words. List five scope drivers and how each drives pay.
CONFIRMATION: For each driver, ask me three questions, one at a time, to size true decision authority, budget responsibility, and organizational impact before finalizing.

dollars. Simply feed AI the job description of your choosing and ask it to extract scope elements such as budget size, team size, decision rights, number of cross-functional partners, and portfolio complexity. Then, compare that to the BLS (Bureau of Labor Statistics) role families and pay bands. Next, adjust your target ask based on structural scope, not title inflation. Remember, if your scope is bigger than the posted range, you aren't being "difficult." You're asking for accuracy and what you deserve.

Outcomes: Bring Receipts

I always encourage my clients to keep a "Heck, yeah!" file of measurable outcomes. This is simply a log of all your wins, both big and small, and the documented impact they have. To do this, tie each outcome (win) to one of four buckets: revenue generated, cost saved, risk reduced, or capability built. Then, match each outcome to an executive-friendly metric and a stakeholder who would vouch for it. Close the loop with a short sentence that connects your outcome to the business priority. When you track the outcomes, you're making your value visible. Leaders who document progress are consistently more likely to advance. Your visibility compounds this opportunity.

Scarcity: Name What is Rare About You

A global survey completed for 2025 found that skill gaps are the top barrier to transformation, and leaders plan to pay to compete for scarce talent.[116] Because when the skills are rare, and the stakes are high, the market pays more. Are you bilingual? Do you have global experience? What niche skills do you possess? Use these to name your edge, and then price it accordingly. Scarcity demands a premium, and, girl, you are one of a kind.

PROMPT: OUTCOMES RECEIPTS

When to Use this Prompt: Use this prompt when you're about to ask for more pay, more scope, a new title, or more influence. And you want your case to land as facts, not feelings. It's especially powerful before performance reviews, promotion conversations, or any moment where impact needs to be visible, credible, and impossible to dismiss.

ROLE: You're my executive communications partner who knows how leaders translate impact into influence.

GOAL: Convert the raw wins I paste below into a one-page value dossier designed for negotiation. For each win, clearly connect the outcome to a measurable metric, name the receipt that proves it, identify the stakeholder who benefited, and write a two-sentence business link that explains why this mattered to the company. The result should read like evidence, not just my opinion.

CONTEXT: I will paste unpolished notes, dashboards, emails, and highlights from my work. I need help turning scattered proof into a clear executive-ready narrative that holds up in a comp review, promotion discussion, or compensation negotiation.

CONSTRAINTS: One page. Plain language. No jargon. Prioritize clarity, credibility, and confidence over volume.

CONFIRMATION: Ask me five rapid-fire questions, one at a time, to sharpen metrics, clarify outcomes, and strengthen the proof before finalizing.

PROMPT: SCARCITY SIGNAL

When to Use this Prompt: Use this prompt when you want to stop competing on qualifications and start competing on *irreplaceability*. It's especially powerful before negotiations, executive interviews, or counter-offers, or at any moment where you need to justify *why* you command a premium.

ROLE: You're a world-class market strategist who understands how scarcity, not effort, drives pricing power.
GOAL: Identify three skills I bring that are genuinely scarce relative to this role and translate each into a clear pricing narrative. For every skill, name the capability, anchor it to one concrete metric, and spell out the business risk the company carries if this role remains unfilled or underpowered. The outcome is leverage rooted in market reality.
CONTEXT: I will paste my resume and the job description for the role I'm exploring. I want to understand where my experience creates outsized value and how to articulate that value in language that decision-makers respect.
CONSTRAINTS: 120 words total. One sentence per skill, and one metric. Include one metric and one business risk for each. No fluff and no adjectives without evidence.
CONFIRMATION: Ask two targeted questions, one at a time, to validate true scarcity with proof before finalizing.

Risk: Price the Weight You Carry

If you own legal exposure, revenue targets, on-call crisis, or brand reputation in the wild, that risk is already priced into market medians. Use AI to map your risk profile and compare it to peer roles. As risk goes up, your compensation should move with it. There's a reason people with C-level roles command the highest pay, as they carry the highest level of accountability when things go right as well as when they go wrong.

Leverage: Timing, Alternatives, and Framing

Leverage is about understanding your real options and knowing how to use them. It comes from timing, including where you are in the company's cycle, such as budget planning, performance reviews, or moments when new scope is being defined. It also comes from the alternatives available to you, whether that's another opportunity, a scoped project that demonstrates value, or a clearly defined path to expanded responsibility, supported by strong research and evidence. Research on negotiation shows that ambiguity can increase bias, while clear data, defined scope, and objective criteria make it easier to be heard and reduce the likelihood of backlash.[117]

What Research Says About the "Ask"

The narrative that "women don't negotiate" is outdated in many contexts. Among recent MBA cohorts, women sometimes ask more than men and still face lower outcomes. At the same time, in early-career transitions where information is thin, men are still more likely to negotiate first offers, and both women and men underestimate the gains from asking. The common thread is information. When standards are clear and the range is visible, results improve.[118]

PROMPT: RISK PROFILE TO PAY

When to Use this Prompt: Use this prompt when your responsibility has quietly expanded but your compensation remains the same. It's especially powerful before race conversations, scope resets, or promotion discussions, or at any moment where increased risk needs to be made visible and priced appropriately.

ROLE: You're a compensation risk analyst who evaluates pay through the lens of decision authority, downside exposure, and organizational accountability.

GOAL: Translate the responsibilities I paste below into a clear risk profile and a recommended pay adjustment. Identify where I personally absorb risk and explain how that exposure should be reflected in compensation. The outcome is a grounded, executive-ready rationale for why my pay should move with my risk.

CONTEXT: I will paste my current responsibilities, decision rights, and performance metrics. I want to understand how much risk I'm carrying relative to my title and pay band, and how to articulate that imbalance clearly and credibly.

CONSTRAINTS: Keep it to 150 words maximum. Name the top three risk exposures and, for each, explain the logic for the proposed compensation adjustment in plain language suitable for senior leadership.

CONFIRMATION: Ask me three questions, one at a time, to help you get to 85% certainty.

PROMPT: THE FRAMING SHEET

When to Use this Prompt: Use this prompt right before the conversation that matters, when preparation is done and it's time to speak. It's ideal for live negotiations, counter-offers, or promotion discussions where you need to lead with clarity, not caveats, and anchor the room before anyone else does.

ROLE: You're my negotiation coach focused on clarity, power, and composure under pressure.
GOAL: Create a 60-second framing statement that clearly connects my market data role scope and documented outcomes to one specific compensation ask. Then generate three calibrated trade-offs I can offer without weakening my position, plus one line that explicitly names the benefit to the stakeholder making the decision. The result should sound steady, informed, and hard to argue with.
CONTEXT: I will paste my market brief scope profile and outcomes dossier. I need a concise, executive-ready frame I can deliver live without over-explaining.
CONSTRAINTS: Keep the framing statement under 60 seconds when spoken, clear, direct, and free of filler or over-explanation. Use confident, neutral language with no hedging, disclaimers, or apologetic tone. Anchor the statement in specific outcomes, scope, and market alignment, not effort or potential.
CONFIRMATION: Interview me by asking four questions, one at a time, to confirm the exact compensation asked, my strongest non-negotiable, which trade-offs are truly acceptable, and who the final decision-maker is before drafting the framing statement.

Backlash is more likely when expectations are ambiguous. Research by Hannah Riley Bowles and colleagues shows that gender effects are strongest in those moments when evaluators rely on assumptions instead of criteria. Reduce ambiguity, anchor to clear standards, and align your language to shared goals. That approach keeps the conversation focused.[119]

Pay transparency moves the needle by making ranges visible and forcing internal discipline. States continue to adopt salary range posting laws, and HR leaders report growing adoption of pay equity analytics to detect and correct these gaps.[120] This is a meaningful step in the right direction, and one that creates visibility and pressure for change.

AI as Your Financial Ally

In compensation systems shaped by data, financial modeling becomes a protective mechanism. Market benchmarks are your starting line. Once you know the range, use AI to keep your value visible, your numbers current and impactful, and your message crisp. If you're using AI in a negotiation for a new position (either internally or externally), use it to model total compensation like a Chief Financial Officer.

Use AI to compare base salary, bonus, equity, sign-on 401(k) match, benefits, and remote or travel costs. Ask for side-by-side tables if you have more than one opportunity on the table, break-even timelines, and "what would need to be true" options.

A "*what would need to be true*" option is a strategic decision-making tool used by CFOs, consultants, and executive leaders to evaluate possibilities without getting stuck in a "yes/no" pattern of thinking. In this scenario, I want you to pivot from thinking, "I want more money" to "For this role, here is what would need to be true..." You can ask AI to build a "what would need to be true" scenario for this job offer. "What adjustments would need to be made to make this a top-tier competitive package?" From there, you can work with AI to generate a "hell-yes" scenario versus a minimum-acceptable scenario. It gives you a clear map of exactly what to say and why before you ever walk into the room.

Once you have your "hell-yes" scenario, you rehearse your ask until you feel confident. Don't let your voice waver. I want you to activate voice diction mode and ask AI to: "Play the role of a financially conservative hiring manager who's responsible for negotiating my salary for this new role. Interview me in real time as if we're in an actual compensation discussion. Ask me the hard questions and push back on my number. I want you to challenge my reasoning." Then, after each response, give me direct feedback on my delivery. I want to know what worked, what weakened my position, and what I should tighten to advocate for myself more powerfully."

Practice until you feel confident, refine your tone as needed, and anchor to the data, one question at a time.

Note this, too. You can use AI as your financial ally in more than just salary negotiations. In HER Collective, we use AI as a "Career Equity Tracker," and the members use AI to track projects they have completed, professional wins, skills acquired, and revenue saved and generated. Members use AI to run an impact report, allowing them to quickly pull together quarterly value summaries for both themselves and their team members, and year-over-year career wins.

You can, and should, also use AI as a CFO for your career, helping you model offers, trade-offs, and long-term value, not just the number on the offer letter. This tool is useful when you're comparing more than one opportunity or weighing a current role against a new one with real trade-offs.

PROMPT: LIVE ROLE-PLAY

When to Use this Prompt: Use this prompt when the numbers are real and the stakes are high, such as before final interviews, offer negotiations, or internal comp discussions. It's designed to help you practice holding your ground calmly, clearly, and credibly before you're in the room where it counts.

ROLE: You're a financially conservative hiring manager responsible for protecting budget, internal equity, and precedent. You're supportive but skeptical, and you negotiate firmly.

GOAL: Simulate a realistic compensation conversation. Push back on my ask, test my reasoning, and help me strengthen my confidence, clarity, and delivery. After each response, provide direct feedback so I can refine how I can advocate for myself.

CONTEXT: I will paste my framing sheet, including my proposed compensation range, market data, and "hell-yes" scenario. This conversation should mirror a real negotiation. Polite, professional, and high stakes. Assume I want this role, but I'm advocating for a package aligned with my value and the market.

CONSTRAINTS: Ask up to eight questions in total, one at a time. Push back realistically on compensation, scope, and timing, and after each response, give me one specific note on what worked, one clear language tweak to strengthen my position, and one data point or reinforcement that I could use more effectively. Don't soften pushback or "help me win" during the role play.

CONFIRMATION: Ask me five questions, one at a time, to clarify the key details needed before drafting.

PROMPT: THE STRATEGIC OFFER EVALUATOR

When to Use this Prompt: Use this when you're evaluating a new job offer, promotion, or compensation change and want clarity beyond the headline number. This is especially powerful when two options feel close on paper, when equity or bonuses complicate the picture, or when you need to understand what actually builds long-term value instead of short-term validation.

ROLE: You're the CFO for my career and focused on long-term value, risk, and sustainability, not just short-term cash.
GOAL: Help me evaluate and compare job offers by modeling total compensation over three years. Identify break-even points and highlight which levers matter most for my financial and personal goals.
CONTEXT: I will paste the details of each offer, including:

- Base salary
- Bonus
- Equity
- Benefits
- Flexibility
- Any current costs or trade-offs

I want to understand how each option performs over time, not just on paper today.
CONSTRAINTS: Create one clear side-by-side table and include a three-year projection and break-even analysis. Be sure to call out the two compensation levers with the highest return on investment and provide a concise summary (120 words max) explaining which option best aligns with my goals and why.
CONFIRMATION: Interview me by asking four questions, one at a time, to confirm risk tolerance, equity preferences, time horizon, and non-cash priorities before finalizing.

A New Foundation for Wealth

When women have access to clear data, the negotiation and financial landscapes change. AI offers structure when there was ambiguity, evidence where there were assumptions, and strategy where there was emotional labor. It pulls the conversation out of the shadows of guesswork and into the realm of real, measurable value. This is about economic agency for women and economic power. This is about using technology to make the pay gaps even more visible, the motherhood penalty more provable, and making your contributions undeniable. Compensation is the entry point. Capital is the compounding layer. From this clarity, a new question emerges. "*What becomes possible when women are equipped not just to earn the same amount of money as men, but to build something that lasts?*" That's the next frontier. Not just income or one-time negotiations, but generational wealth. Because the real shift isn't just earning more; it's building assets, equity, and influence that extend beyond your role and beyond your lifetime. Wealth works the same way performance systems do: it compounds over time. Wealth compounds through the systems you build and the ones you participate in. The structures you participate in today will determine the value you can accumulate tomorrow. And in an economy increasingly shaped by automation and capital concentration, compounding is not neutral. AI will not correct inequity in compensation systems. It will scale the logic that those systems already reward.

TLDR

- The pay gap and motherhood penalty are real. Women, especially mothers, are systemically underpaid and undervalued, which makes data-backed advocacy non-negotiable.

- AI is the financial ally that gives you clarity, data, and leverage.

- Benchmark your worth with real numbers. AI can pull market medians, salary ranges, and credible comparisons fast.

- Translate your work into real dollars by quantifying your impact and negotiating from evidence, not emotion.

- Use AI to script and rehearse your ask. Preparation is leverage!

YOUR AI-POWERED ACTIONS

ASK:

When was the last time you asked for what you actually wanted (or more!) and believed you deserved it? What story are you telling yourself about money and worth?

APPLY:

Use AI to benchmark your range from official sources, your scope profile, and your value stack. Draft your framing sheet and rehearse with a live role-play.

AMPLIFY:

Teach other women how to research their market value. Share what you have learned, compare notes, and make the invisible visible. When more women know their value, it raises the floor for all women.

CHAPTER 8

BUILDING GENERATIONAL WEALTH

I spend a lot of time thinking about what I want my kids to inherit. But I'm not talking about dollars or assets. I'm talking about possibilities. I want them to grow up in a world where their choices aren't limited by money, exhaustion, or the fear of starting over. I want them to feel the kind of freedom most women spend their entire lives trying to earn.

But that kind of financial freedom does not happen by accident or luck. It happens because someone, somewhere in the family (or a really incredible friend) decided to build more than a paycheck. They built stability. They built options. They built space to breathe. And what they were really building, whether they called it that or not, was a system.

Wealth isn't built by intention alone. It's built by systems that compound what they are designed to compound.

That's what I'm trying to do in my own life, and in my work with women. I want to create a foundation strong enough that the next generation doesn't have to fight the same battles. AI won't make you wealthy on its own, but it can make wealth-building simpler to start, easier to sustain, and harder to avoid. It turns the abstract dream of "financial freedom" into something visible, trackable, and possible, one decision at a time. And when you zoom out beyond your own balance sheet, you start to see the pattern.

The Wealth Reality Without the Fluff

If wealth compounds through systems, then unequal systems compound unequal outcomes. The data shows a fact that isn't surprising in the least: women hold less wealth than men do. When the Federal Reserve looked at household wealth in 2022, it found a significant difference. In households where men answered the survey, the typical amount of wealth was represented as $1. In households where women answered, the typical amount of wealth was only 55 cents. This means that, on average, households led or represented by women hold about half as much total wealth as those represented by men.[121]

Wealth includes things like savings, home value, retirement accounts, investments, business ownership, and anything that adds to your net worth. The finding is simple: even today, women on average

"Wealth isn't built by intention alone. It's built by systems that compound what they are designed to compound."

have built far less long-term financial security than men.[122] While the gap is not as wide as it used to be, it's still a major sign that economic equality is unfinished work.

The picture changes when you zoom in. Among single adults, the typical man had $82,100 in wealth in 2022, compared with $58,100 for the typical woman.[123] Single women without children were closer to parity, which reminds us how caregiving, career interruptions, and policy design shape outcomes as much as personal behavior.[124]

Income is a piece of the puzzle. Mothers who work full-time, year-round, still face a sizable pay penalty relative to fathers. Recent analyses put moms at about 74 to 88 cents to the father's dollar, depending on the measure used, and the gap widens when part-time work is included.[125] Pay influences what women can put away, and what they can put away shapes the wealth they are able to build over time.

Retirement data tells the same truth from another angle. Vanguard's 2025 *How America Saves* shows persistent differences in account balances by gender, driven largely by income and tenure.[126] Plan design features like automatic enrollment, automatic increases, and target date funds help close that gap. They lift outcomes for everyone, but they matter even more for women who have less margin for error and fewer years of uninterrupted compounding to rely on.

And all of this is unfolding at the exact moment trillions of dollars are changing hands. Economists call it the Great Wealth Transfer.

The Great Wealth Transfer is the movement of assets from Baby Boomers and the Silent Generation to their children, grandchildren, and chosen causes. Between now and the mid-2040s, an estimated $84 trillion to $124 trillion will transfer hands in the United States alone.[127] The question isn't whether this wealth will move; it's whether it will compound inside redesigned systems or reinforce the ones that already exclude women. Homes, retirement accounts, investments, businesses, cash, land, life insurance—all of it will move. In addition, women are more often named as trustees, executors, and primary beneficiaries.[128] Research indicates that daughters today are increasingly named direct beneficiaries in wealth transfers.[129] These are not just assets that shape a life, but assets that can reshape a family for generations.

But here's the part we cannot overlook. Wealth moves through systems as much as it moves through families. Generational wealth is built through compounding advantages, and access to the tools that accelerate compounding isn't distributed equally.

The tools that now shape how wealth compounds are increasingly algorithmic, which means access to AI literacy is becoming access to leverage. Women still lag in AI adoption and formal AI education. Only about one-third of surveyed U.S. women report using generative AI, compared with 44% of men.[130] When capital allocation systems increasingly rely on AI-informed analysis, participation gaps quietly become wealth gaps. Among those who do use AI, women are less likely to use it daily and often lack structured training opportunities.

If wealth compounds through leverage, and AI increasingly shapes leverage, then skill gaps quietly become wealth gaps. Building wealth in the AI age means investing in your skills, not just your savings, so you capture the upside of this technological shift instead of financing someone else's.

Not only that, but women are positioned to control more of this wealth than ever before because they are also **building it**. Women are out-earning previous generations.[131] They are starting businesses at record rates.[132] Women are becoming executives, founders, advisors, and investors. Women are generating wealth through leadership, ownership, expertise, and grit, not just through family transfers.

We also tend to live longer, outlive our male partners, and we increasingly manage family finances.[133] This isn't a small opportunity but a complete shift in financial power.

I always say you have to play the game to win the game—and to change the game. This is where we change the game. Some women will inherit wealth. Some will build it. Many women will do both, and I'm here to support each scenario. Whether the money comes from a parent or a paycheck, the same truth applies—receiving wealth and growing wealth are two different skill sets. One is passive, the other is intentional. And the difference is preparation. Preparation is what turns income into leverage. It's what transforms opportunity into optionality. Because the woman who's financially prepared does not make decisions from fear or dependence. She makes them from alignment. And that shift changes everything. **Optionality only matters if you can act on it.**

F*ck You Money: The Real Beginning of Freedom

Before we talk about engineering wealth, we need to talk about *optionality.* When I use the phrase "F*ck You Money," I mean **walk-away money**. It's having enough cash set aside that if your values are violated, your work is dismissed, or your pay no longer reflects your worth, you can leave without panic, without scrambling, or without renegotiating

"If wealth compounds through leverage, and AI increasingly shapes leverage, then skill gaps quietly become wealth gaps. Building wealth in the AI age means investing in your skills, not just your savings, so you capture the upside of this technological shift instead of financing someone else's."

your dignity. You don't need permission. You don't need approval. And you don't need to stay.

Financial independence is structural leverage, and I can't stress enough how important this is for your overall freedom, stability, and peace of mind. It changes how you move, what you accept, and how you advocate for yourself. That is why the idea of F*ck You Money resonates, because it represents real freedom. It gives you the ability to choose your life without the fear of what will happen if you say the honest thing, take the harder path, or refuse to stay in environments that diminish you, especially the toxic, dismissive, mansplaining ones so many of us know so well.

When I was climbing in the corporate world, my partner and I made the intentional decision to save. It was the quiet discipline of a woman who understood that my future may require courage, and courage tends to arrive more fully when supported by a financial cushion. I always knew there could come a day when the room I was in no longer aligned with the woman I was becoming.

And when that moment came, there was no panic or frantic math about "can we afford this?" There was no sacrificing my values because I needed a paycheck to provide for my family. There was only clarity. I didn't have to shrink myself to make anyone else comfortable or tolerate disrespect and misguided visions. Because I had saved on purpose, I had options.

So, when the final line was crossed, I didn't negotiate with myself. I simply walked away. And I did it with my voice and my integrity intact. That's the real meaning of F*ck You Money.

Optionality is built through a financial architecture that runs long before you need it. If you take nothing else away from this story, take this: **give your future self the freedom to walk when walking is the most honest thing you can do**. Everything we build from here is designed to create and protect that freedom. The systems, automations, and AI give you the support to grow steadily and build that kind of freedom. The slow, steady, powerful kind that lets you say, when necessary, "I'm not tolerating this anymore. F*ck You."

There's one thing I need to say clearly: F*ck You Money has to be built long before you need it. You don't decide to walk away and then figure out the money. By the time you're being disrespected, undervalued, or asked to compromise your values, it's already too late to start building leverage. In those moments, panic replaces clarity, and fear becomes a negotiating partner. F*ck You Money only works if it's already there.

That's what we're engineering next. Ideally, you never need it, but just in case you do. F*ck You Money comes from a quiet system that's built when things are stable. The hard part isn't the idea (everyone wants a pile of cash in savings!). It's the consistency. The math. The discipline to keep going when nothing feels urgent—*yet.*

This is where AI is genuinely useful. It can automate the boring parts, remove guesswork, and turn your walk-away plan into a simple engine that runs quietly in the background. AI helps you decide how much to build, how it grows, and how to check your readiness, so you're never making this decision under pressure or fear. Walk-away power protects you. The wealth engine grows you.

The Wealth Engine You Can Actually Run

I like to think of money as energy. It moves, grows, and flows through your life in seasons. In some seasons, it accelerates; in others, it may consolidate. But it's always in motion. Left unmanaged, that energy can feel chaotic or controlling. **Wealth is what happens when that energy is organized into a system.**

True wealth is the result of a system working quietly in the background, one that channels money with intention instead of reacting to

THE FREEDOM FUND BUILDER

When to Use this Prompt: Use this prompt when you're ready to build your Freedom Fund (aka, F*ck You Money) intentionally, not reactively. It's for the moment you decide that security isn't about having a job, but about having options, even if you never use them.

ROLE: You're a certified financial-literacy coach and calm, practical operator. You prioritize sustainability over intensity and clarity over perfection.

GOAL: Design my Freedom Fund plan (walk-away money), so I can build a financial buffer before I ever need it. Create a plan with:

- A realistic monthly savings target.
- A step-up schedule as income grows.
- Automated transfers that fit my real cash flow.
- A one-page Walk-Away Readiness Check I can review quarterly.

CONTEXT: I will paste my details below. My goal is not deprivation; it's optionality and peace of mind.

- Monthly take-home pay: [INSERT]
- Fixed bills: [INSERT]
- Average card spend: [INSERT]
- Current savings: [INSERT]
- Income stability pattern (steady/fluctuating/seasonal): [INSERT]
- Desired confidence level for walk-away (low/medium/high): [INSERT]

CONSTRAINTS: Plain language. 250 words max. Use ranges, not absolutes. Prioritize automation over willpower. Your output must include:

- A recommended monthly Freedom Fund contribution with a conservative starting range.
- A simple step-up plan tied to raises, bonuses, or income increases.
- Where to keep the fund (account types, not specific products).
- How to label transfers in my bank so the purpose stays visible.

- A short three-sentence script I can use to say no to new expenses while the fund is being built.
- And a quarterly Walk-Away Readiness Check with three to four indicators that show progress.

CONFIRMATION: Begin by asking up to five questions, one at a time, to clarify:

- What truly counts as essential spending.
- My comfort with short-term trade-offs.
- The earliest date I'd want the option to walk away.
- How predictable my income really is.
- What usually derails my saving consistently.

it. And like any system, it reinforces what it's built to support. If money is energy, then wealth is the engine, one you can actually run when you click these five gears into place. Each gear reinforces the others. Remove one, and the system strains. Strengthen all five, and the system starts to work for you.

1. **Earn** more with clarity and negotiation.
2. **Keep** more with automation and tax-smart decisions.
3. **Grow** more with steady investing aligned to risk and goals.
4. **Protect** more with buffers, insurance, and good paperwork.
5. **Teach** what you learn so money wisdom outlives you.

AI lives under each gear and acts as your Financial Chief of Staff, gathering data, spotting patterns, and protecting the rhythm so you can keep your money goals on track.

Gear 1: Earn—Your Paycheck Funds the Plan

You did the Value Equation work in Chapter 7. Keep this momentum going. Market ranges, scoped responsibilities, and receipts translate into a higher base, bonus, and benefits. The closer you get to objective standards, the more bias loses its leverage.

Your AI Move: Keep a running "Career Equity Tracker" that logs revenue generated, costs saved, risks reduced, and capabilities built. Each quarter, ask your AI assistant to format a one-page impact memo you can paste into performance systems or investor updates. This makes raises and equity grants less personal and emotional and more professional. It also gives you the clarity and confidence to ask for your worth.

Gear 2: Keep—Automation Beats Willpower Every Time

Real financial stability comes from reducing the guesswork. When you automate the movement of your money (what goes into savings, what you invest, debt paydown, etc.), you remove the daily guesswork that derails consistency. Automation replaces willpower with structure, removes the mental load, keeps you consistent, and protects your goals.

Defaults are your friend, so **auto-save and auto-invest** wherever possible. Vanguard's plan-design research shows that automatic enrollment in your 401(k), automatic escalation, and age-appropriate target-date funds or model portfolios increase participation and keep portfolios aligned to long-term goals.[134]

Once your savings are on autopilot, the next step is deciding what happens to the money once it leaves your checking account. If you invest through a 401(k), an IRA, a robo-advisor, or your own brokerage, the goal isn't just to contribute dollars. It's to make sure the portfolio you are building behaves the way you want it to. This is where **automated rebalancing** comes in.

Automated rebalancing is the quiet maintenance system inside your investments. Over time, some parts of your portfolio may grow faster than others. If stocks have a strong year, suddenly you may be holding more risk than you originally intended. If bonds grow, you might be holding more safety than you planned. Rebalancing simply resets the mix back to the allocation you chose at the start. Automation allows you to maintain the level of risk you desire. Rebalancing adjusts your portfolio on a schedule, removing the guesswork, with no emotional decisions or headaches.

If you're looking for more information on automation, reach out to the firm that handles your finances, or explore platforms like Ellevest. I met Sallie Krawcheck, founder and former CEO, when I interviewed her on stage at a women's conference and instantly connected with what she'd built. Ellevest is a wealth management and financial planning services company that's optimized for women. Ellevest was built specifically to account for women's different realities, and you can find more information at www.ellevest.com. Just remember, you don't need a big chunk of money to get started. You only need consistency over time.

A word of caution: many "smart" savings apps help you move money, but they may have monthly fees that can eat into a small balance. Use them only if they simplify actions that you wouldn't take on your own.

Your AI Move: Build a monthly Money Pulse that pulls your transactions, classifies patterns, and surfaces two decisions: where more automation would help, and where a spending pattern should be adjusted because it's no longer aligned with your goals. Those small recalibrations add up.

Gear 3: Grow—Invest Steadily and On Purpose

Women are investing at higher rates than a year ago, yet many still report lower confidence and higher stress.[135] Fidelity's 2024 Women & Investing study found 71% of women now own stock market investments, which has increased sharply from 2023. While the "confidence gap" remains, women are nearly twice as likely as men to describe their investing knowledge as "non-existent."[136]

The good news is that progress and pressure can coexist. The answer is education that women can turn into action. Progress in financial literacy is real, but so is the pressure women feel to "get it right." Remember that perfection isn't the answer. You just need to learn enough to take the next best step, and then let your action build momentum.

If you want values to guide your dollars, the marketplace is catching up. Sustainable and impact options are more available than ever, but they aren't always straightforward. "Sustainable" typically means your money is invested in companies that aim to operate responsibly across environmental, social, and governance factors. "Impact" goes a step further, focusing on investments designed to create measurable positive outcomes, like clean energy, healthcare access, or advancing women in leadership.

"The good news is that progress and pressure can coexist. The answer is education that women can turn into action. "

The challenge is that these labels are not standardized. What qualifies as "sustainable" or "impact" can vary widely depending on the fund, and disclosures aren't always easy to interpret. In some cases, the language is clearer than the actual results.

Before you commit, take a closer look at how a fund defines impact, how it screens companies, and what fees are tied to that strategy. The OECD's consumer guidance offers a helpful starting point for understanding your rights, risks, and what transparency should look like in sustainable finance.[137]

Your AI Move: Ask your AI assistant to translate the complexity for you, summarize a fund's purpose, fees, historical volatility, and top holdings in plain English. (I like to say, "Explain this to me like you would a 5th grader.) If you use a robo-advisor, download its Form ADV or disclosures and ask AI to explain how it profiles risk, rebalances your portfolio, and manages cash, so that you have working knowledge of how the system is working behind the scenes.[138]

One last thing: if this section feels confusing or overwhelming, keep going. This isn't a sign that you don't belong. Every CEO, every millionaire, every person who now speaks this language with ease once sat exactly where you are. No one starts financially fluent. Everyone who understands this now once had to learn it. This is where it starts.

Gear 4: Protect—Safeguard Your Plan from Life

An emergency fund turns a crisis, sudden job loss, or financial dry period into a problem you can solve. It removes some of the stress and gives you space to think clearly instead of reacting under pressure. Insurance and essential estate documents protect the people you love from the financial consequences of your absence or incapacity, and with tens of trillions of dollars moving to heirs and charities over the next two decades, preparation matters more than ever. Families who plan ahead will have fewer conflicts and better tax outcomes. This gear is all about ensuring you have the documents you need in place: a simple will, updated beneficiary designations, and a shared document vault can reshape the outcome dramatically.

Your AI Move: Ask your AI assistant to build a list of all essential documents you should have, where each one will live, who has access, and which beneficiaries need reviewing. Have it generate short scripts to start the conversations that matter with your partner, parents, or your kids, so the emotional weight of these topics feel more manageable.

Gear 5: Teach—Multiply What You Know

Money habits are taught in the kitchens more than they are in the classrooms. That's why generational wealth doesn't start with assets, but with conversations. If you want generational wealth, you have to teach generational wisdom.

In my house, that looks like my daughter Halle's business, **Sibling Sparkle**, and she runs it with the confidence only a kid with a dream and some glittery disco ball beads can have. And because she's watching me build a company, she wanted to build one too. Halle started this business when we started dragging her to her brother's travel baseball games. She had the incredible idea to create and sell the trendy necklaces that are popular with baseball players. The idea was perfect. It would give her something to do during the long hours at the field, and she could make money (one of her favorite pastimes).

We have "investor meetings" where we sit down and talk about the cost of beads, review current inventory, and review the previous weekend's revenue. We talk about profit margins, pricing, and product ideas. She taps into her brother and his friends for her ideal client feedback. I ask her what she's saving for, and she tells me how she wants to reinvest in new materials. It isn't perfect, but it's adorable and completely intentional. These moments where we talk about revenue, costs, and the lesson about paying yourself first are the seeds of generational wealth for Halle. Not because she's making thousands of dollars, but because of the mindset and confidence it's building within her.

Your AI Move: Use AI to create a "Family Financial Debrief." This is a tight, 30-minute meeting that you can run weekly or monthly. Ask it to produce three elements: one insight (something learned about money, choices, or an insight into the results), one adjustment (something to change), and one alignment (a decision or value you will bring into the next week). AI can tailor it to your kids' ages, your household, and your financial goals. The point of this exercise is to get you, and the family, into a rhythm. When you normalize money conversations, you normalize financial literacy and power. This is how kids gain the confidence to build something long before they ever need to inherit it. And when systems grow stronger, the guardrails matter just as much as the gains.

Your Safety is Part of the Strategy

AI can sharpen your financial decisions, but it should never replace your judgment. Keep these safety rules in mind if you are using AI to empower your financial decisions.

YOUR 10-SECOND SAFETY CHECK

Before you use AI for anything related to your finances, run this quick scan:

Who built this tool? If it's not from a regulated institution or a reputable provider, use caution.
What information am I about to share? Never share Social Security Numbers, full account numbers, or deeply personal identifiers.
Where is this data going? Look for a clear privacy policy and data handling statement before connecting accounts.
Why is this recommendation being made? Notice if the platform is nudging you toward a product it profits from.
How big is this decision? For small choices, AI is fine, but for high-stakes financial moves, cross-check with a financial expert.

Use Regulated Providers

If you're paying for financial advice, human or digital, verify the firm's registration on the SEC's Investment Advisor Public Disclosure site. Review their Form ADV, the official disclosure document that outlines the firm's services, fees, and any conflicts of interest. Check the fee schedule and scan for anything that could influence their recommendations.

Read the Fine Print

The Consumer Financial Protection Bureau (CFPB) warns that some digital platforms use subtle design tricks to push you toward products that benefit the company more than the consumer. Make sure you're comparing offers outside of the app. Read the fine print. If something feels "too guided," do your due diligence.[139]

Always Protect Your Privacy

Never enter Social Security Numbers, full account information, or deeply personal details into general chat tools. Use institution-provided AI features if they are available, and read the data policy before connecting your financial accounts.

Engineered for the Freedom to Walk Away

Generational wealth is built through early, quiet, and consistent decisions that are made long before they feel urgent. Systems compound what they are designed to compound. It's built when things are steady, when you still have choices, and when the math can be done calmly. It lives in systems that run quietly, decisions made early, and habits that don't depend on mood or urgency.

When you engineer F*ck You Money, you give yourself leverage, you give yourself time, and you give yourself the ability to leave a situation that no longer fits without bargaining away your values or your voice. That freedom doesn't show up all at once; it compounds. It's preparedness. It's knowing that your life, your work, and your family aren't trapped by a single paycheck or a single season.

When designed with intention, wealth becomes a stabilizing force that supports choice, resilience, and dignity. This is how freedom becomes structural instead of situational. And in systems increasingly shaped by automation, what compounds will depend on what we choose to build.

PERFECT PROMPTS TO PUT THIS CHAPTER TO WORK

1) Freedom Fund Architect

If the Freedom Fund was about building your first layer of walk-away power, this next prompt goes deeper. This isn't about saving toward a number. It's about redesigning how your money flows, so optionality is built into your system every month. The Freedom Fund gets you breathing room. The Architect builds the structure that keeps you free long-term.

Use this prompt when you're ready to move beyond building a safety net and begin designing a money system that supports stability, investing, and optionality automatically.

ROLE: You're a values-aligned financial planner and cash flow strategist focused on building optionality and walk-away power over time. You think in systems, not spreadsheets. Your job is to design money flows that quietly build stability, protect choice, and support long-term freedom without relying on constant attention, discipline, or sacrifice. You prioritize early preparation over reactive decisions, automation over willpower, and sustainability over intensity.

GOAL: Design a values-aligned cash flow system that quietly builds optionality over time. The goal is to make my money predictable, supportive, and boring in a good way so I can fund stability today and walk-away power tomorrow. Create a one-page plan that:

- Directs money first to essentials, then to freedom, savings, and walk-away money, investing, and giving.
- Uses target ranges instead of rigid limits so the system can adapt as my life changes.
- Automates savings and investing so progress continues even when motivation is low, reduces decision fatigue, and removes the risk of waiting until a moment of crisis to act.

CONTEXT: I will paste the last 60 days of transactions by category and my top three values. My income may be steady, fluctuating, or seasonal, and my goal is to design a system that works across real life. I'm building this budget in advance while things are stable, so my money supports long-term optionality and the ability to walk away if needed. I want a plan that reduces decision fatigue, replaces willpower with structure, and quietly supports wealth building over time.

CONSTRAINTS: One page. Plain language. Uses ranges, not rigid limits. Clearly separate fixed vs. variable income. Prioritize automation over manual tracking. Suggest a safe automation date based on my pay cycle. Design for sustainability, not short-term intensity. Include a "stop doing" recommendation that reduces friction and frees up cash. Avoid guilt-based framing, moralizing, or budgeting jargon.

CONFIRMATION: Before drafting, ask up to five questions, one at a time, to clarify:

1. **Essentials:** What expenses truly need to be protected no matter what, and which ones feel essential out of habit, not necessity?
2. **Volatility:** Where does my income or spending fluctuate most, and how often does that create stress or surprises?
3. **Freedom priority:** What would building walk-away money make possible for me in the next 12–24 months?
4. **Friction:** What consistently breaks my saving or budgeting intentions when life gets busy or emotional?
5. **Stability signal:** What would a "good month" with money look like if this system were working as designed?

Begin by asking the first question.

2) Five-Year Freedom Modeler

Use this when you want to see where your money is actually taking you. This is especially powerful before changing jobs, adjusting investing contributions, accelerating debt payoff, or planning a major life move, when small shifts today could dramatically alter your freedom five years from now.

ROLE: You're my personal CFO focused on durability, optionality, and long-term freedom.
GOAL: Build a five-year financial projection that shows how my savings, debt payoff, and investing contributions evolve under three scenarios: conservative, steady, and stretch. Make the trade-offs visible so I can see how small decisions today compound into real leverage over time. The outcome is clarity about which path buys me the most freedom with the least regret.
CONTEXT: I will paste current balances, interest rates, minimums, retirement contributions, and take-home pay. I want a grounded, realistic model that helps me make confident choices even when life changes.
CONSTRAINTS: No market predictions. Use a range of historical real-return assumptions and show them clearly. Include an action list with dates I can paste into my calendar.
CONFIRMATION: Interview me by asking six questions, one at a time, to confirm: my emergency fund target, employer match details, upcoming life events, tolerance for variability, and any constraints that matter before producing the model.

TLDR

- Wealth isn't just earned; it's engineered with systems that run on autopilot.

- The gender wealth gap is still real, with survey data showing that households with women respondents hold about 55 cents of wealth for every dollar held by households with male respondents.

- Women are investing more, but confidence and access still lag. Clear education, simple systems, and automation shrink that confidence gap dramatically.

- The Great Wealth Transfer is a once-in-a-generation window. Tens of trillions of dollars will move to heirs and charities over the next two decades.

- AI is your Financial Chief of Staff. Use it for clarity, pattern-spotting, automation, and decision support. Let AI gather the data, simplify the options, and surface questions.

- Always refer to humans, advisors, attorneys, and financial planners for judgment, advice, and complexity.

YOUR AI-POWERED ACTIONS

ASK:

What does wealth mean to me beyond a number? What would freedom look and feel like at home, at work, and for the next generation?

APPLY:

Use the Freedom Fund Architect to engineer F*ck You Money before you ever need to use it.

AMPLIFY:

Share one financial lesson with a friend, or with your kids. Visibility around money creates generational wealth. Normalize having conversations about money, where you're investing, how much you're making, and how you plan on growing your money.

PART 4

CLOSING THE LEADERSHIP GAP

Defining What Lasts

By this point, you have changed how you see the system, how you move inside it, and how you build leverage beyond it. The final layer asks a deeper question: what gets locked in?

AI accelerates tasks and carries forward the patterns it learns. The behaviors organizations reward, the language leaders use, and the criteria that shape promotion and compensation become part of the architecture automated systems learn from and reinforce. Over time, those patterns solidify into standards. Standards become defaults. Defaults become infrastructure.

This is the part of the Leadership Gap most people never see. We often think of leadership as how we show up in a moment. But at scale, leadership becomes infrastructure. When that infrastructure reflects outdated or incomplete models, the gap is reinforced at scale.

The Infrastructure Gap is the distance between participating in a system and shaping what comes next. It is the space between adapting to existing norms and redefining them before they harden into code. When leadership models fail to evolve, automation preserves them. When leadership expands, systems expand with it.

Closing this gap requires leadership that is intentional enough to scale. It asks you to think beyond individual advancement and toward institutional impact. The conversations you normalize, the behaviors you reward, the communities you build, and the standards you model don't stop with you. They carry forward.

Infrastructure is built quietly through repeated behavior, consistent language, and shared expectations. When you lead with clarity and build deliberately, those patterns extend far beyond you.

In this section, you will learn how to lead in a way that scales, how to anchor your leadership in your values, judgment, and voice, how to expand your impact through community, and how to step into the next great leap as a leader shaping what comes next.

AI is already reshaping the future of work, so the question is: which leadership model will scale alongside it?

The answer is already taking shape. It lives in how you lead, what you normalize, and what your leadership sets in motion.

CHAPTER 9

LEADING LIKE YOU

As AI becomes embedded into organizational workflows, leadership itself becomes infrastructure, and whatever model of leadership we fail to redesign will be codified into the systems that scale. The magnitude of that shift is easy to underestimate, which makes the next question unavoidable.

What if leadership in the age of AI isn't about outsmarting the machines or how AI-fluent you are, but about out-human-ing the people who rely on them? We've all heard the tired warning that "AI will take your job," but that's never been the real threat. It's not AI that replaces people. It's people who know how to use AI who outperform those who don't. If we go deeper, the ones who will succeed and rise the highest aren't just technically fluent; they're the ones who pair that fluency with the very skills AI can't replace: empathy, intuition, connection, emotional insight, and the ability to read a room without a single word being spoken. If we don't consciously define that leadership model, AI systems will default to amplifying the dominant archetypes already embedded in organizational data. The question is no longer who performs best, but which definition of leadership gets hard-coded into the architecture of decision-making.

So, ask yourself this: what does leadership look like when the person using AI isn't just efficient, but emotionally intelligent, deeply self-aware, and fully authentic? That's the leadership model that will shape how AI systems behave inside organizations. For decades, organizational leadership standards privileged dominance, certainty, and visible authority. Yet the capabilities least susceptible to automation, such as judgment, relational intelligence, and interpretive awareness, have historically been undervalued in formal leadership models. As AI systems scale decision-making, those previously discounted strengths become central to how decisions are made and how organizations operate. The shift isn't accidental. It reflects a realignment between what organizations once rewarded and what automated systems cannot replicate.

Women, who have long developed these skills in environments that demanded social fluency and adaptive awareness, are positioned at the center of that realignment. If leadership norms are shifting structurally, then individual leaders must participate in that redesign deliberately.

This chapter invites you to rethink leadership entirely.

I don't want you to think of it as performance that you manage, or a standard you chase, but a way of showing up that's consistent and unmistakably yours. It's about leading from a place that isn't curated by the people before you, but that's grounded in your actual values, your actual voice, and your actual instincts.

Redefining leadership isn't a cute branding exercise or about finding a new framework. It's a structural safeguard against cementing outdated power models into automated systems. The role of AI isn't here so you can produce the world's perfect leadership announcement that sounds nothing like you. It's to clear out the unnecessary noise and friction, so you have the space, clarity, and confidence to lead as the most honest version of yourself. This is where authenticity becomes strategy and humanity becomes the advantage. This is where you discover the kind of leader that only YOU can be, and exactly why in this era of AI, it's your time to shine. And when you zoom out, that shift is already visible across the data.

The Moment Women Were Built For

Before going further, let's dive into what the data tells us about leadership in the age of AI. Across industries, the skills least likely to be automated sit squarely in women's wheelhouse: empathy, active listening, judgment, and values-aligned influence. The World Economic Forum's 2025 Skills Outlook lists empathy and active listening among today's core skills and notes that they face very low substitution by current generative AI.

Leadership and social influence are also rising in importance as work becomes more interdependent and change accelerates. As organizations layer AI into daily operations, fewer decisions are purely technical, and more require coordination, judgment, and alignment across teams. Work now moves through networks, not silos, which means leaders must translate strategy across functions, manage ambiguity in real time, and influence without relying solely on authority.

Organizations increasingly rely on leaders who can coordinate across functions, make decisions under uncertainty, and build trust in environments shaped by rapid technological shifts. These capabilities carry strategic value because they determine how quickly ideas move, how confidently teams adapt, and whether innovation actually sticks.[140]

At the same time, rigorous field experiments show that access to generative AI measurably boosts worker productivity, especially for less experienced employees, by simplifying routine cognitive work.[141] These gains scale execution, yet leadership remains essential for framing problems, setting standards, and connecting work to purpose.

Women are already leading effectively in this phase of work, one defined by sustained disruption, frequent crisis, and constant change. During periods of uncertainty, women leaders have been rated more

---

"Redefining leadership isn't a cute branding exercise or about finding a new framework. It's a structural safeguard against cementing outdated power models into automated systems."

effective across a wide set of competencies, including inspiring motivation, developing others, and building relationships.[142]

Evidence from 2020 through 2024 shows women leaders consistently receive higher effectiveness ratings in areas that stabilize teams during disruption, including collaboration, communication, and relationship-building. These capabilities function as anchors during volatility, supporting trust and continuity when systems are under pressure.[143] Those patterns show up in the most human moments of work.

Around the time of writing, I heard a conversation where someone wondered out loud if AI might someday replace call center agents entirely. While I hear that fear often, and on paper it sounds efficient, I spoke with a woman who works in life insurance, and she said something that stopped me in my tracks. "*We would never automate those calls*," she said. When I asked her to tell me more, she informed me that those calls aren't about policy numbers or data verification; they're about people navigating one of the worst moments of their lives, the death of someone they love. In those moments, a script can't hold space for the grief that often can't be held back, and even the best bot can't read between the lines or pick up on what isn't being said. No algorithm can replace the steady human presence that says, "I'm here with you. I got you."

That changed the way I viewed AI capabilities and how far they could go. While AI may streamline the work and make it easier to find answers, there are places where our humanity and compassion can't be replicated. That boundary is becoming more clear as AI systems move deeper into organizational life.

As generative AI adoption accelerates and the gender gap narrows in mainstream tools, leadership strengths grounded in judgment, connection, and trust increasingly shape outcomes at scale.[144] In this era, leadership capacity is defined by the ability to pair technological fluency with human judgment, and the skills women already embody align directly with that demand.

What makes these strengths even more powerful in an AI-shaped workplace is authenticity. Research across sectors shows that authentic leadership—leaders whose actions align with stated values—correlates with stronger employee trust, higher organizational commitment, and greater discretionary effort.[145] Similar findings appear in interdisciplinary studies linking authentic leadership to improved employee well-being, psychological safety, and resilience during change.[146] When you place these findings side by side, a clear pattern surfaces.

Taken together, the evidence points in one direction. AI can scale your capabilities, but your humanity is what creates the conditions for people and businesses to thrive. It shapes trust, drives connection, and determines whether the systems being built actually work for the people inside them. The leaders who succeed in this era are the ones who pair technological fluency with emotional intelligence, authenticity, and courage. These aren't new skills for women. They are strengths we've been building all along.

History shows us something about systems: they reward what they were designed to recognize. For decades, leadership models elevated dominance, certainty, and command-and-control authority. Those traits were treated as neutral, but they were calibrated around a narrow archetype. As AI becomes embedded into organizations, those calibrations matter. If we don't question the assumptions beneath our leadership models, technology will simply scale them. AI will not soften a narrow definition of power. It will reinforce it.

Expanding leadership now prevents hard-coding yesterday's model into tomorrow's systems. Leading like you requires widening the definition before the system hardens it into default. And that requires structure. That structure must be practical, repeatable, and durable enough to shape behavior before it becomes the system norm.

Leading Like You: A Simple Operating System

If leadership becomes infrastructure in the AI era, then it must be designed with the same intentionality as any other operating system. If this moment in leadership rewards judgment, trust, and human connection, then the real question becomes practical. *How do you lead this way consistently, especially as the pace of work accelerates and AI becomes embedded in everything you do?*

The research tells us these are the strengths that matter now. Yet living them consistently is different from simply knowing they matter. There's a moment in every woman's leadership journey when she realizes she's done performing, shapeshifting, and contorting herself into whatever version of "professional" the room once demanded. The realization isn't a failure of ambition. It's a signal that the old models no longer fit the work being asked of you or the leader you have become.

In an era where AI can replicate efficiency but not integrity, this moment matters. What follows is a way to operationalize what the data already confirms: that your judgment, your voice, and your humanity are the most powerful strategic assets you have.

Think of what follows as a **Leadership OS (Operating System)**. It's a repeatable governance layer that stabilizes judgment, decision-making, and communication before those patterns are absorbed into automated systems. It consists of five repeatable practices that are light enough to use daily but durable enough to shape how leadership shows up at scale.

These practices include:

- **Values to Voice:** Turn your values into decisions.
- **Meeting to Meaning:** Clarify decisions and next steps.
- **The Empathy Pre-Brief:** Enter hard conversations grounded.
- **A Bias Check for Decisions:** Widen perspectives before committing.

- **A Reflection Rhythm:** Notice patterns and adjust intentionally.

As AI becomes embedded in every workflow and decision, this operating system becomes the tether that keeps you grounded so you continue leading like *you.*

Values to Voice

Every meaningful leadership moment comes down to this: what do you stand for when the stakes get real? In *Glass Ceilings and Sticky Floors,* I cover the importance of knowing your personal core values. And I'm not talking about your company's core values posted on the hallway walls, but you as an individual. What do you care about? What matters most to you? Understanding what your core values are is the first step.

Next, you have to translate those values into words that your team can feel. This is about bringing your values out of the clouds and into the room. Before any board call, high-stakes meeting, or tough conversation, ask your AI to help you turn your values into behavioral guardrails. Behavioral guardrails are the lived expression of your values. It's the way that your principles become visible, audible, and felt by the people you lead. They can help you answer questions like:

- How do I make decisions when two values compete?
- What do I refuse to compromise on?
- What do people experience when they encounter my leadership?

This takes the guesswork out of who you need to be in the toughest moments. They quiet the self-doubt that flares when the stakes rise. When you know your values and have strong guardrails, you no longer have to recalibrate your integrity or spend precious time making the right decision. You become a leader people can predict and trust.

PROMPT: VALUES TO VOICE

When to Use this Prompt: Use this prompt when your values are about to be tested by speed, pressure, politics, or power. It's designed for moments when *how* you speak matters as much as *what* you decide, and you want your voice to sound steady, ethical, and unmistakably yours.

ROLE: You're my executive communications partner and ethics spotter, helping me hold my values when the room gets tense.
GOAL: Translate my three core values into clear decision guardrails and ready-to-use language for an upcoming call. For each value, define what I will do, what I will say, and what I will not do when those values are in tension. The outcome is language that feels grounded, human, and principled. But please, no corporate buzzwords.
CONTEXT: I'm navigating a real decision involving specific stakeholders. I want to lead with clarity and consistency, even if the choice is uncomfortable or unpopular.
CONSTRAINTS: 150 words. Plain language. Map each guardrail to one likely risk.
CONFIRMATION: Interview me by asking three questions, one at a time, to confirm my core values, non-negotiable red lines, and who needs to hear this message before drafting.

Meeting to Meaning

You know those meetings that feel like time sucks? Where everyone is talking but no one is landing, and you leave with more questions than you started with? This practice transforms your agenda into a clean two-column snapshot of Decisions Made and Next Moves that are tiny, doable, and have assigned owners. Bonus points if you use an AI tool like Otter or Granola to take meeting notes and optimize next steps.

The Empathy Pre-Brief

At some point in their leadership life, every leader has to have a hard conversation they don't want to, but absolutely must. This could be letting someone know they didn't get the promotion they hoped for, that their project proposal was rejected, or that they were impacted by the most recent reduction in force and no longer have a job as of today. Those moments are some of the hardest in leadership. There's an anxiety that hangs in the air and a tension that sits in your chest. You often worry that you may not say the right thing, or the impact you WANT to have doesn't land. This is where the Empathy Pre-Brief comes in.

I prefer voice dictation on this one, but you feed AI the context (what the topic is, the relationship, and any other information that might be helpful) and ask it to reflect back a one-minute briefing that helps you enter the conversation grounded. Ask it to surface the other person's likely concerns, name the shared interest that you can lead from, and generate one validating sentence that starts with "I can see why..." or "It makes sense that..."

And because you're still you, you bring the humanity—the compassion. AI simply clears the emotional static so you can lead the way and take the leap.

PROMPT: MEETING TO MEANING

When to Use this Prompt: Use this when a meeting ends with a lot of discussion but not enough direction. If you find yourself wondering who owns what, what happens next, or how to move forward without rehashing the conversation, this prompt turns that moment into clear, actionable progress so work can move forward without unnecessary follow-ups, confusion, or quiet resentment. AI handles the translation and administrative work so you can focus on leading.

ROLE: You're the chief of staff for my brain. You listen for decisions, accountability, and momentum. You reduce cognitive load by turning conversation into clear direction and forward motion.

GOAL: Transform meeting notes into a decision-ready record and a simple action plan that moves work forward. Produce a clear decision log with a two-column action plan organized by owner, due date, and effort estimate. Identify one risk and one dependency per action. The goal is shared clarity, not perfect documentation.

CONTEXT: I will paste raw meeting notes or a transcript below. I lead a hybrid team and work across functions. I value momentum, clear ownership, and starting with small, doable next steps rather than large value stacks.

CONSTRAINTS: Paste-ready bullets. Suggest calendar holds. Draft one short update I can send to stakeholders to confirm alignment and prioritize 10-minute first steps wherever possible.

CONFIRMATION: Interview me by asking up to three questions, one at a time, to confirm final owners' true deadlines (not aspirational ones) and scope boundaries to prevent work creep before finalizing.

PROMPT: EMPATHY PRE-BRIEF

When to Use this Prompt: Use this prompt right before a conversation where tone will matter as much as content—feedback, disagreement, alignment resets, or boundary-setting with someone you need to keep working with. It's designed to help you walk in calm, credible, and hard to misread.

ROLE: You're my coach for candid, high-stakes conversations. You're grounded, perceptive, and emotionally fluent.
GOAL: Create a one-minute pre-brief that helps me enter a difficult or sensitive conversation steady and prepared. Identify likely concerns on both sides, name shared interests, and draft two validating openers that lower defensiveness without diluting my point. Include one well-placed question that invites voice and signals respect, plus a clear boundary I can use if the conversation drifts or escalates.
CONTEXT: The conversation is about [topic] with [person/level]. Our relationship is [brief description], and I want to communicate clearly while protecting trust and momentum.
CONSTRAINTS: No therapy language. Clear, adult, direct. Add one boundary if the conversation derails.
CONFIRMATION: Interview me by asking two questions, one at a time, to clarify their incentives and mine before drafting.

A Bias Check for Decisions

Every leader has blind spots, and the smartest leaders KNOW they do. This practice slows the moment long enough for you to see what speed would have overlooked: your confirmation bias, the sunk-cost trap, the halo-effect around a high-performer, or the tendency to overpressure women and underpressure men (which happens more often than most leaders would like to admit).

A bias check uses AI as a neutral auditor to pressure test your leadership and to widen the frame. Here's how I use it. I paste in the decision I *think* is the decision I want to go with, and I ask AI to tell me what risks I may be overlooking, the alternative framings I didn't consider, or the piece(s) of data I need before I commit to my decision. There's no removal of judgment, just the widening of the possibilities, but with attuned risk.

A Reflection Rhythm

One of the most impactful lessons I've learned in leadership is that reflection is the key to making future decisions better. Reflection is where you can recalibrate your leadership by noticing your patterns, where you can tune in to where you may be drifting from your desired goals, and where you can reclaim the pieces of yourself that may get scattered during a hectic career. This practice keeps your self-awareness fresh.

Each week, you can either speak to your AI or copy/paste a journal entry in and ask it to give you a pattern readout on: where your energy spiked, where it drained, where your impact landed, where your boundaries may have slipped, and what micro-shift you should take next. I used to think this was a "weird" practice. Talking to AI about my deepest fears and concerns initially felt like I was trying to make AI my therapist, but then I realized it's about stewardship of your energy, your presence, and your leadership identity.

PROMPT: BIAS CHECK FOR DECISIONS

When to Use this Prompt: Use this prompt when a decision feels obvious, urgent, or unanimously supported—that's usually when bias is loudest. It's especially powerful before high-visibility calls, people decisions, or irreversible moves where fairness, credibility, and long-term trust matter more than being fast.

ROLE: You're a neutral decision auditor with no stake in the outcome and a sharp eye for power, pattern, and blind spots.
GOAL: Audit the decision I paste below for cognitive and organizational bias. Name the specific bias at play, explain how it could distort judgment in this situation, and propose a concrete counter-move to reduce its impact. For each bias, identify one missing data point or dissenting perspective I should actively seek before locking the decision. The outcome is a clearer, fairer, more defensible choice.
CONTEXT: I will paste a brief decision summary and list the stakeholders involved.
CONSTRAINTS: 200 words. Clearly name each bias and its fix. Use plain language.
CONFIRMATION: Ask three questions, one at a time, that could challenge my current view.

PROMPT: REFLECTION RHYTHM

When to Use this Prompt: Use this when your weeks blur together and you sense patterns forming but haven't paused long enough to name them. This rhythm is especially powerful during seasons of high visibility or sustained output, when small energy drains can quietly compound into burnout if left unexamined.

ROLE: You're a burnout-prevention coach and mirror.
GOAL: Generate a weekly five-question voice journal. Compile a monthly pattern summary with energy spikes, drains, impact receipts, and one shift to test.
CONTEXT: Fast-moving leadership role. Hybrid team. High public visibility.
CONSTRAINTS: Five questions max. Less than five minutes. Simple language. Keep it to three to five micro-shifts I can make next month.
CONFIRMATION: Ask two questions to tailor prompts to my season.

The Leadership OS at Work

If we put these five steps into play, you might use your Values to Voice before a big stakeholder meeting to frame trade-offs with clarity and courage. Then, after a tense escalation, you run the Empathy Pre-Brief before the follow-up call. You open the conversation with validation, lead with shared purpose, and name a boundary that the team needs to adopt with calm precision. Then, if the plan starts to feel unbalanced, you whip out the Bias Check to surface any dissenting data that was missing. To protect your edge, you close out each Friday afternoon with a Reflection Rhythm, letting your pattern report become the dashboard that you trust more than your high-functioning brain.

Before Leadership Becomes Default

Leadership norms carry forward as technology advances and becomes embedded. The behaviors organizations reward, the language managers use, and the signals that earn promotion are the very patterns AI systems learn to recognize and reinforce. If performative leadership remains the dominant model, automated systems will amplify it. If relational, judgment-based leadership becomes the norm, those patterns will scale instead.

For decades, women were asked to adjust themselves to fit leadership environments calibrated around narrow archetypes. In the AI era, the more consequential question isn't whether women can adapt to those models, but whether organizations will redefine the model before it becomes permanently structured into automated decision-making.

AI doesn't determine the future of leadership, but it does operationalize whatever definition we normalize. The opportunity in this moment is to influence structure through how you lead.

"AI will make you faster, more productive, and give you a new level of clarity you haven't been able to experience yet."

When you use AI to quiet the noise, shrink the administrative clutter, and minimize the mental load, what rises to the surface is the leader you have always been underneath the pressure of "doing all the things." You create space to be the leader who notices the energy in the room before anyone says a word. The leader who can hear what isn't being said. The leader who asks better questions because she actually cares about the people around her. These aren't soft skills; they are leadership skills, and they are about to become the most valuable skills of the next era.

AI will make you faster, more productive, and give you a new level of clarity you haven't been able to experience yet. It will help you identify patterns, rehearse tough conversations, and make more data-driven decisions. But it cannot make you more human. That's all you, and that's your advantage.

The Real Advantage

Where I really anchored my power was in realizing that I didn't have to be deeply technical to lead in the age of AI. In fact, I don't identify as deeply technical, and I don't come from a coding or computing background. What mattered far more was my willingness to show up fully to bring my judgment, my voice, and my humanity into the room. That realization changed how I led, and it clarified where my real leverage lived.

When AI absorbs administrative load and pattern recognition work, what remains most consequential is judgment. The real work is defining the leadership standards that guide those tools. If those standards aren't deliberate, automation will normalize whatever behaviors and archetypes are already embedded in organizational data. Leadership in the AI era centers on defining the model before it scales. What gets built next reflects how you choose to lead now.

TLDR

- AI doesn't invent or adjust leadership norms to the latest fad. It scales the ones already operating inside organizations, and those systems are failing women and other marginalized groups.

- Performance standards, promotion criteria, evaluation language, and compensation logic function as upstream infrastructure.

- If those systems privilege narrow archetypes, automation will intensify them.

- If organizations redefine leadership around judgment, relational intelligence, and accountability, those are the patterns that will scale.

- The question isn't whether AI will change leadership. It's which definition of leadership will become the default.

YOUR AI-POWERED ACTIONS

ASK:

Where in my leadership am I reinforcing a model I wouldn't want automated and scaled? What behaviors, language patterns, or decision defaults might become the template AI systems learn from?

APPLY:

Select one upcoming decision, meeting, or performance evaluation and deliberately redefine the leadership standard you want to normalize. Use one of the prompts in this chapter not for self-optimization, but to clarify the model you are reinforcing.

AMPLIFY:

Name the leadership behavior you want to become standard inside your organization. Reinforce it publicly, reward it visibly, and model it consistently. Systems scale what becomes normal.

CHAPTER 10

THE POWER OF COMMUNITY

I've always believed in the power of community. I've seen how much farther women can go when they are supported by other women who understand the weight they carry and the ambition they hold. The women who go the farthest don't go alone. They go **with a circle**—a small, trusted group that offers perspective, accountability, shared wisdom, and support along the way.

That circle doesn't slow them down or confine them. It steadies them. It makes growth sustainable instead of lonely. Community is infrastructure. It redistributes access, shortens learning curves, and compounds opportunity.

This chapter is about the power of **rising together**. About how belonging, shared knowledge, and collective intelligence accelerate growth in ways no single woman could ever replicate on her own. If the last decade has taught us that visibility matters, the next decade is teaching us that community turns visibility into impact. And in a world reshaped by AI, where speed, complexity, and constant change are the norm, community is no longer a "nice to have" for women leaders. It's part of the infrastructure that allows them to learn faster, adapt more confidently, and step fully into their potential.

HER Collective: Community and AI, Working Together

Like it or not, we're entering an era where humans working with AI together will outperform every other model. You've seen what AI can do when you're using it alone. What's far more powerful and far more sustainable is what happens when women learn, test, and apply AI *in community.* As AI accelerates individual output, community becomes the structural counterweight that protects judgment from narrowing and power from consolidating.

HER Collective was designed with that belief at its core. From the beginning, this wasn't meant to be a place where women simply gathered to talk about leadership or ambition. HER Collective was designed to compress the distance between ambition and opportunity. It functions as a shared environment where women learn how to use AI in ways that support their leadership, careers, and lives, without needing to be experts first and without carrying the learning alone. That support is embedded directly into the experience through specialized AI agents designed to accelerate career growth and decision-making, giving women practical tools to translate ambition into action.

Without shared environments of learning and calibration, AI risks amplifying isolation rather than expanding opportunity. Community transforms AI from an individual productivity tool into a shared learning infrastructure.

Inside HER Collective, AI is woven into how women think, reset, and move forward over time. One example is the **HER Vision Architect**—a custom GPT available to members whenever they need to step back and reorient. Women use it to clarify their direction at the start of the year, to re-prioritize after a season of burnout, or to regain momentum when life or leadership shifts unexpectedly. It's a space to articulate what they're building, what they're projecting, and what needs to change.

We also have **Camille, the Career Strategy Architect**. She is a one-, three-, and five-year career planning agent that helps you map your strategy forward. Camille helps you see clearly where you are now, define where you want to go, and model your options. Do you stay or go? Take a promotion or a lateral leap? Women use Camille before asking for a raise, applying for a promotion, during burnout, before a big leap, and any time they feel stuck between comfort and ambition.

Another shared tool is **Melinda,** a custom-built resume enhancer and career positioning GPT inside HER Collective, designed to help women translate their experience into outcome-focused, KPI-driven language. Melinda is optimized for applicant tracking systems, removes "soft" language and replaces it with gender-neutral language, and pushes women to focus on specific KPI-focused outcomes. Women use Melinda to clarify impact, surface scope, and articulate results in ways that align with modern hiring systems and leadership expectations.

"Without shared environments of learning and calibration, AI risks amplifying isolation rather than expanding opportunity."

What gives these tools their power is how they're used together. Women share how they're applying AI, refine language collaboratively, and compare approaches. One member experiments with Sheryl, my digital-twin executive coach GPT inside HER Collective, to practice boundary-setting. Another uses Melinda to prepare for negotiations and reports back. Someone else revisits her Vision Architect mid-year to reset her direction, while someone else uses Camille to focus on strengthening the skills she needs for her *next* role and plots how to acquire them.

One woman's experiment becomes another woman's shortcut, and momentum accelerates. This creates collective intelligence in motion. AI accelerates clarity through shared tools and a shared language. The community strengthens judgment, accountability, and confidence through conversation and support. Over time, women build fluency in AI alongside confidence in how they lead and make decisions. What happens inside a small circle mirrors what happens inside institutions. Access compounds through connection.

This aligns closely with what research has shown for years. Studies of MBA graduates found that women with strong networks paired with a small trusted inner circle of other women were significantly more likely to advance to senior leadership roles. The difference wasn't about their capabilities but the connections they had.[147] Institutions reward proximity to power. Networks create it. Research on belonging shows similar effects on performance, resilience, and retention.[148] Together, these findings reinforce what many women already know: that leadership accelerates when it's shared.

HER Collective exists to put that insight into practice by combining community, shared accountability, and AI as a practical partner in growth. Without deliberate structures of collaboration, AI will optimize individual output while quietly eroding collective influence.

AI makes it easy to work alone. You can put your headphones on, open your laptop, and spend hours thinking, drafting, deciding, and refining without ever checking your assumptions out loud. It feels focused. It feels productive. But isolation has always been a structural disadvantage for women. The right community creates leverage.

And if you're not paying attention, it can slowly shrink your world down to just you and your own thinking. The best leaders know that a variety of perspectives and lived experiences lead to better decisions and stronger outcomes. When ideas are challenged, refined, and expanded in conversation, outcomes improve. Community brings exactly that. It creates space for different viewpoints, shared wisdom, and real-time feedback that strengthens judgment and confidence. In an era shaped by AI, those perspectives matter *so much*. Yes, we can move fast on our own, but as my favorite Peloton instructors always say, "Together, we go far." Leadership in the age of AI is no different. Access to collective intelligence isn't a luxury in a systems shift, but it's a requirement for shaping it.

When You Don't Have a HER Collective (Yet)

If you're reading this and thinking that a community like HER Collective sounds incredible but won't work for you right now, you can start smaller. Think of three to five women in your world who are ambitious, honest, and looking to grow. They might be at your current company, a former company from your network, from your neighborhood, or from your life outside of work.

Reach out and tell them the truth: "I refuse to do this alone anymore. I want to build something that feels like a shared lab for our careers, our leadership, and our growth. In partnership with AI, would you be open to trying this with me for three months?" Find a meeting cadence that works for you, and create a shared agenda that includes:

- One win she is proud of
- One challenge she needs help with
- One way she's experimenting with AI

You can use a simple shared document or a group chat to drop prompts, scripts, or different resources. If one woman tries a new AI tool to help with time blocking or salary research and it works, she can share her process. All of a sudden, everyone in the chat gets smarter. The point isn't to recreate HER Collective, but to stop doing it alone. Over time, what we build together becomes the environment that shapes what's possible for all of us.

Community as Part of Your Leadership Legacy

The famous Maya Angelou quote, "People will forget what you said, people will forget what you did, but people will never forget how you made them feel," captures what comes next. This is where community becomes part of your leadership legacy.

At this stage in your career, your success is no longer defined by individual metrics, milestones, or the bullets on your resume. Success

"HER Collective exists to put that insight into practice by combining community, shared accountability, and AI as a practical partner in growth. Without deliberate structures of collaboration, AI will optimize individual output while quietly eroding collective influence."

is defined by how many women rise because you existed. Movements are built when individual success becomes a collective standard. By how many women feel seen, supported, steadied, and strengthened simply by being in your presence. When you invest in a community, you expand your leadership. You're choosing to be the woman who doesn't pull up the ladder behind her, but instead, turns around and reaches a hand back to lift up other women, while climbing the ladder herself. Believe me when I say, there's more room up the ladder than anyone ever wanted us to believe.

AI will keep evolving, and the tools will keep changing, but the thing that will never change is who you're becoming and whom you bring with you. HER Collective is one expression of that commitment in my world. It's my way of showing women that we don't have to white-knuckle this by ourselves anymore. We can bring our ambition and our exhaustion and still rise. And whether you step inside HER Collective or not, I want you to feel this at your core. You were never meant to carry your career, your family, your leadership, your AI literacy, or your life entirely on your own. The future of leadership will be built by women who architect ecosystems—women who rise together, share leverage, and refuse isolation as the default. In an era where AI scales, whatever structures we normalize, the communities we build become part of the architecture of the future.

TLDR

- Community isn't optional anymore. It's your competitive advantage. Find yours.

- Women rise higher, faster, and healthier when they don't rise alone.

- Belonging, shared knowledge, and collective intelligence accelerate growth in ways no single woman can recreate by herself.

- AI will absolutely make you more effective on your own, but the real power comes when women use it together. One woman's prompt becomes another woman's shortcut.

- When women have space to be fully themselves, practice power without fear or punishment, to try, fail, and learn again without shame, everything changes. Confidence compounds, careers accelerate, and possibilities become infinite.

YOUR AI-POWERED ACTIONS

ASK:

Where am I still trying to do this alone? Where might I find a group of women who can help me shoulder my ambition, my exhaustion, and my curiosity about AI?

APPLY:

Find a community like HER Collective or choose three to five women who are hungry for more and invite them into a three-month AI and leadership circle. If you already have a HER Collective, recommit to showing up as you are and bring a friend!

AMPLIFY:

Share what you're learning with other women. Talk openly about the role community and AI are playing in your growth and leadership.

CHAPTER 11

THE NEXT GREAT LEAP FORWARD

The choices we make during moments of technological change ripple far beyond our own careers. When my kids are my age, I don't want them inheriting a story of gaps. I don't want to have to explain why their mother's generation was still making less money for the same work, still out-credentialed and under-promoted, still doing three shifts a day while the systems around us were insisting the playing field was "mostly fair."

I want to be able to look them in the eye and say, "We caught it in time. We saw the next wave coming, and instead of getting swept under, we learned how to ride it." Moments like this reshape institutions, incentives, and access across generations. The wave is here. It's AI. And the question in front of us is not "Will AI take our jobs?" The real question is: will women use this moment to close every gap we've been carrying, or will we let those gaps get coded permanently into the future of work?

Here, we stop treating AI like a threat or a tool on the side and start seeing it as a once-in-a-generation opening. AI isn't a threat waiting in the wings. It's a structural redesign already underway, and it's moving at a rapid pace. Everything you've built—your clarity, your receipts, your leadership operating system, your financial leverage—now becomes directional. This is where it all turns outward.

This is the leap. Not just for "women in tech," but for all women. Full stop.

This is our time. And if we're going to claim this moment, we need to understand the ground we're standing on.

The World You Are Walking Into

I'm going to ask you to zoom out with me for a moment. The World Economic Forum's Global Gender Gap Report estimates that at our current pace, it will take roughly 135 years to fully close the global economic participation gap between men and women.[149] This gap measures how equally men and women are able to participate in, and benefit from, the economy and includes:

- **Labor force participation:** Are women working at the same rates as men?

- **Wage equality for similar work:** When a woman does the same job as a man, is she paid the same?
- **Income levels:** Are women earning comparable lifetime income?
- **Leadership representation:** How many women hold senior executive and managerial roles?
- **Access to advancement:** How easily can women move up, gain promotions, or shift into high-earning fields?

These barriers are structural. They include persistent pay gaps, the "motherhood penalty," underrepresentation in executive roles, overrepresentation in lower-paying or unpaid labor, gendered expectations around caregiving, slow progress in STEM representation, and AI reshaping jobs faster than women are being included in the decisions.

At the same time, work itself is being rewritten. The World Economic Forum's Future of Jobs analysis found that over the next five years, through about 2030, 170 million new jobs are projected to be created globally while 92 million existing roles are likely to be displaced or fundamentally transformed.[150] That's a positive gain of about 78 million jobs globally. This is a great reshuffling at massive scale.

The same report estimates that about 22% of jobs worldwide will experience structural labor market churn in the next five years. That means roughly one in five roles will change in ways that affect job content, skills required, or even whether that role exists at all. Employers also expect that nearly 39% of the skills they consider "core" today will change by 2030.[151] That means:

- Almost four in 10 skills you rely on today will not be the skills that carry you through the next decade.
- About one in five jobs will be reshaped, relocated, or retired.
- New roles will emerge faster than old ones disappear, but they will not be evenly distributed.

This isn't just disruption but a complete re-architecture of the workforce, and it will redesign how everyone will do business moving forward. At this scale, shifts in work inevitably reshape how value is defined and who holds authority. Whenever work gets rebuilt at this scale, power gets rebuilt with it. If women aren't in the rooms where new roles are defined, if we're not the ones experimenting with the tools, setting guardrails, and deciding what "good work" looks like in an AI-powered world, we risk something worse than being left behind.

We risk watching the old gaps get poured into the concrete of the new economy. The systems will still be built. They will simply be built without us. We have seen this pattern before in other systems that were built without women at the center.

History shows what happens when women's needs are treated as an afterthought. Women are 50% more likely to be misdiagnosed after a heart attack, and 17% more likely to be killed in a car crash because safety standards were built around male bodies.[152, 153] Systems designed for the "reference man" quietly scale harm. AI follows the same pattern. If we allow today's tools to be trained on yesterday's defaults, we will

"This isn't just disruption but a complete re-architecture of the workforce, and it will redesign how everyone will do business moving forward."

hard-code inequality into the architecture of the future. The workplace is simply the next system being redesigned under pressure.

What's Really at Stake for Women

You already know that women are still underrepresented in leadership and overrepresented in roles that are highly exposed to automation. That reality hasn't changed. What is changing is the speed and the scale.

A recent policy brief from the OECD, "Algorithm and Eve: How AI will Impact Women at Work," found that AI users and companies are more likely to be younger males and highly educated. Women, especially those in clerical administrative roles, are less likely to report using AI at work and more likely to be sitting in jobs that are structurally vulnerable to automation.[154]

At the same time, a report from the International Labour Organization flagged that AI is poised to transform a far higher share of female-dominated occupations than male-dominated ones, particularly in high-income countries where women cluster in office, secretarial, and administrative work.[155] This means the jobs most likely to be reshaped by AI are jobs where women are highly represented, and the people most likely to be learning AI the fastest inside organizations are not women. That's the AI Gap in action. And structural gaps, unlike fixed traits, respond to pressure and participation.

The good news is that the gap isn't fixed. Deloitte's projections showed that women's experimentation with generative AI in the United States has been growing significantly faster than men's. They projected that by the end of 2025, women's usage would meet or exceed men's, even as trust and representation gaps remained.[156] As of early 2026, the gap is closing, but it still exists across the broader market.

Adoption is only the starting point. It doesn't determine outcomes.

The habits, assumptions, and leadership standards that accompany adoption determine whether AI compresses gaps or compounds them. When participation expands without thoughtful redesign, inequities will scale. When participation is paired with intentional leadership, system patterns will shift.

If we want this moment to change outcomes, leadership has to be intentional. That means understanding how AI actually behaves inside systems of power. It amplifies speed, visibility, and scale. Whatever habits, assumptions, and structures are already in place will move faster and spread wider.

Speed magnifies everything. When systems move faster, the consequences of our design choices show up sooner and hit harder. Understanding how AI amplifies what already exists is how we make sure growth expands opportunity instead of quietly hardening old gaps.

AI as an Accelerator, Not a Replacement

One of my favorite studies on AI comes from a large field experiment with 5,172 customer support agents. When some agents were given access to generative AI assistance that could suggest replies and surface best practices, their productivity jumped about 14–15%. The biggest gains weren't for the experts, but for the novice and lower-skilled workers whose performance moved much closer to that of the top performers.[157] This study is a powerful example of AI being a floor-raiser. It's lifting the baseline faster for the people who are willing to lean in and learn. It helps you move out of the weeds so that you can spend more time on the work that only you can do. But it doesn't decide what matters. It doesn't set the vision, define the values, or choose what "good" looks like for a team. It doesn't notice that the brilliant woman on your staff has been quietly carrying the emotional load for the entire group or that your customers are grieving when they call, not just confused about a login. That's the human work. The pattern is clear. AI raises output, but it doesn't decide what actually matters.

Earlier in this book, we named the Five Cs of Humanity as your un-automatable edge: critical thinking, curiosity, collaboration, communication, and connection. AI can mimic your language and write really good emails, but it cannot care about its people. It can decode patterns and identify new ones, but it cannot hold space for someone's fear. It can predict the word that comes next with ease, but it cannot tell you the right thing to say or do at this very moment. The leaders who will succeed in the next decade are the ones who treat AI as an amplifier of their humanity. These are the leaders who let the machines handle the routine, and the humans handle the remarkable.

Women's Great Leap Forward

Let's take a quick look back in history. In the last century, women fought for basic legal and social rights to vote, to own property, to work, to open bank accounts, and not to be punished for getting pregnant.

In the last few decades, we pushed into higher education, professional careers, and leadership roles at rates that would have been unimaginable to our grandmothers. We're now the majority of degree holders in many countries. We're founders, we're C-suite leaders, we're investors and board members. We are glass ceiling smashers.

And yet the gaps have lingered in pay, in promotion, and in representation. In safety, in our health, and in our wealth. The AI era isn't just "another technology wave." It's the first wave we're entering with women already at the table in the lab, on the code repositories, on the cap tables, and in communities where adoption is being shaped. That means this time we aren't begging for a late invitation to a room that was never built with us in mind. We get to decide what this room looks like. This is our great leap forward, and it's already underway.

This moment will not happen by accident, though. It will be built by women who trust themselves enough to lead and learn out loud so others can see them. Women who use AI to multiply their impact, refuse to negotiate their value without receipts or data, who build enough financial runway to walk away from what no longer serves them, to lead like themselves, and to rise together in communities.

You have spent this entire book becoming that woman. Chapter by chapter, you've been quietly engineering your own great leap forward. And here is the part you may not have realized until now: that leap has already begun, not in some distant future version of you, but in the small decisions you've made while reading, reflecting, and applying what you've learned.

The leap rarely arrives as a loud declaration. It builds in the questions you dared to ask, the assumptions you chose to examine, and the habits you began to rewrite. It shows up in the boundaries you practiced, even when your voice shook. It takes shape the moment you choose courage over conditioning and step toward technology instead of turning away from it.

Great leaps rarely feel like leaps when you're living them. They reveal themselves in hindsight, in the steady accumulation of small, defiant choices. Choices made by women who are done waiting for permission and done shrinking to fit a world that was never designed for their brilliance.

You are one of those women. Whether you feel ready or not, something has already shifted. The way you think, the way you decide, the way you lead is no longer the same. And in an era where AI scales the structures we normalize, those shifts extend far beyond you. They extend into the stories the next generation will inherit.

TLDR

- The future of work is being rewritten right now. Between 2025 and 2030, around 170 million new jobs are projected to be created; 92 million roles will be displaced or transformed; 22% of jobs will face structural churn; and nearly 39% of core skills are expected to change.

- AI is eliminating work and redistributing power. If women don't step into AI-augmented roles and decision-making, old gaps in pay, promotion, and representation will harden inside new systems.

- AI is raising the floor. Studies show generative AI can boost worker productivity by roughly 14–15%.

- The skills least likely to be automated are empathy, judgment, active listening, collaboration, and authentic communication—the very skills women have spent a lifetime building.

- Your great leap forward is a series of deliberate choices to learn out loud, architect your career, claim your value, build wealth, lead like yourself, and grow with a community while using AI as your amplifier.

YOUR AI-POWERED ACTIONS

ASK:

When I picture myself five years from now in an AI-powered workplace, who am I in that picture? How am I leaning in and leading? If my daughter, niece, mentee, or younger self were watching my choices right now, what story would she believe about what's possible for her? What do I want that story to be instead?

APPLY:

Choose your leap! Pick one major shift you want to drive in the coming year:

- Becoming AI-confident in your role
- Landing your next promotion
- Negotiating a compensation package that reflects your value
- Growing your "F*ck You Money" and starting to invest
- Joining a community like HER Collective

Use AI to help you map your next three concrete moves toward that goal. Put the first on your calendar and do it this week.

AMPLIFY:

Normalize your use of AI at work. Use AI to help you craft a short, powerful message to your team or leader about how you are using AI responsibly to reduce burnout, increase learning, and make better decisions. At the very least, write a LinkedIn post about it and tag me!

CONCLUSION

THE HUMANITY LOOP

AI doesn't have the ability to change the direction of history on its own. It's a pattern machine. It accelerates the direction already embedded in our systems. AI may have started as a mirror, but you're the one who turned it into a movement.

When you opened this book, AI may have been something distant, something for tech folks, the early adopters, the people who seem unbothered by pushing buttons they didn't fully understand. AI lived out there in labs and headlines, and in the hands of people who never questioned whether they belonged.

But now it's in your hands. It's in your browser tab, your notes app, your drafts, your calendar, your career strategy, your financial models, and your leadership voice.

But most importantly, it's become part of your story. What you practiced in these pages reshapes more than your own trajectory. It influences the systems you now participate in. You didn't just read about the AI Gap. You walked right into the center of it and said, "This won't happen on my watch." You took fear and turned it into questions. Those questions became experiments. Experiments became strategy. Strategy turned into power. And that power is what makes the leap possible. That's the humanity loop in motion.

The more we teach machines about humanity, the more they reveal the patterns we reinforce, consciously or unconsciously. In these pages, you fed AI your ambition, your curiosity, your receipts, your doubts, your values, your dreams, and your truth. You watched it reflect both your brilliance and your blind spots. You saw where you were minimizing yourself, and you saw what you've always been capable of becoming.

And then you took a leap.

A leap shaped by the patterns you chose to reinforce.

And now those patterns are shaping more than you. They are shaping what comes next.

The Loop You've Been Running All Along

This whole journey hasn't been linear, but one that represents a loop of curiosity, creation, compassion, courage, and connection. You've been tracing this loop in every chapter, even if you didn't realize it.

Curiosity became the moment you whispered, "What if?" What if AI isn't here to expose me, but to expand me? What if I'm not behind, but I'm right where I'm supposed to be? What if I can learn this faster than I believed possible? The old narratives loosened, the shame quieted, and your curiosity stepped forward.

Creation followed. You started drafting your visions, prepping for hard conversations, negotiating from power instead of apology, mapping your career, shaping new rhythms, and running deeper numbers on your money. What once felt abstract suddenly had clear form. You built clarity and momentum. You built evidence.

Compassion arrived in the places where your old patterns tried to pull you back: the blank page, the double-checking, the caring for everyone else before caring for yourself. Instead of judging her, you met that version of you with care. You treated hesitation as information and let yourself be a beginner. You started talking to yourself like someone you were responsible for protecting.

And with compassion came deep courage. The kind of courage that lives in the decision to ask for more, to send the email, to counter the offer, to set the boundary, to speak the truth publicly, to leave what was no longer aligned. This didn't mean the absence of fear, but it was proof that you're more committed to your future than to today's comfort.

And then came connection. The moment you realized you weren't meant to do any of this alone. You've found, built or imagined your version of HER Collective. You shared prompts and scripts and data instead of keeping them to yourself. You began speaking honestly about money and ambition and AI and the dreams you once carried in silence. You learned in real time that community is truly about infrastructure and that the right community can be a soft place to land and a firm place to launch.

Curiosity led to creation.

Creation led to compassion.

Compassion fueled courage.

Courage deepened connection.

Connection sparked new curiosity.

And the loop keeps running.

And every time it runs, it leaves a trace in the systems you touch.

The Leap Was Never About AI

And here's the deeper truth: technology has always been a story about power. This book is about power: who holds it, who has been denied it, and what you do with it once it's finally in your hands. AI just happens to be the fastest, loudest, and most concentrated form of power reshaping work right now. It's rewriting job descriptions, redistributing opportunity, shifting which skills matter, and determining who gets seen and who doesn't.

Participation determines how power is distributed. When women engage with AI at scale, visibility shifts, compensation recalibrates, and leadership patterns expand. When women are absent from the rooms where systems are defined, those systems still freeze into place; they simply harden without us. This moment is already reorganizing power. The architecture is being set, and who's inside it will define what comes next. This is where it becomes personal.

For your story.

For your career.

For your money.

For your leadership.

For your community.

The leap was never into the machine. It was into yourself.

You Are Somebody's Turning Point

There's a woman watching you right now. She might be your daughter, your niece, your mentee, or a teammate. She might be the woman in the back row who's never spoken up, but feels braver when you do. She might work three floors below you and only know you by the way you carry yourself in the hallway. She might follow you silently online, saving every post you make.

She's watching how you meet this moment. She's watching how you normalize learning in public. She's watching how you negotiate, how you ask questions, how you share tools instead of hoarding them. She's watching what you treat as possible. The standards you model today become the defaults she will inherit tomorrow.

You're somebody's evidence of what's possible. For generations, women have had to imagine themselves into futures they never saw. We became the firsts, the onlys, the "can she really do that?" prototype. You are changing that. The way you use your voice, your tools, your money, your work, your community, and your values in the age of AI will make it easier for the women behind you to take their power for granted in the best way possible.

Here's the truth that you must never forget: there's no finish line to this work. You will still have days when AI overwhelms you and days when you want to throw your laptop out the window. But none of that means you've failed. It means you're human. And when those moments of overwhelm and doubt come, I want you to come back to the loop.

Let curiosity open the door again. Let creation give your hands something to do and let compassion soften the self-critique inside your head. When you feel safer, reach for connection. Text the woman who gets it and step back into the community. Let yourself be held for a moment so you can continue this work. This work isn't for the faint of heart, but each pass through the loop builds your leadership.

The Last Truth

You might feel fired up right now. You might feel exposed. You might feel both. Good. That means something inside of you is shifting. But this is the last truth I want to offer: you were never the problem. Not your ambition, your sensitivity, your doubts or your desire for more.

The problem was a system that told you to shrink, to be grateful, to wait your turn, and to sit out the transformation defining this century. But you didn't. You picked up this book and stepped directly into the arena. You stepped into a structural inflection point and chose to shape it rather than be shaped by it. You engaged the tools, examined the power beneath them, and began defining the standards that will scale long after this moment passes.

That choice matters because when women like you decide to lead with your full humanity and use AI as your amplifier, everything changes. How companies are structured changes, how wealth is distributed changes, and family dynamics and community all change. So, take a breath and look at where you began. Now look at where you are standing and look at what is suddenly possible.

You don't need permission to continue. You're allowed to be powerful here. You're allowed to be wealthy here. You're allowed to be visible here. You're allowed to be fully human here. You are living proof that the future is female, fearless, and full of infinite possibilities.

I believe at my core that infinite possibilities exist, but they don't appear out of thin air or happen by chance. They are designed. They are built in boardrooms and group chats, in hiring decisions and investment choices, in the courage to ask for more and the refusal to shrink back. AI will keep accelerating whatever surrounds it. What surrounds it is us. The standards we normalize, the power we claim, the communities we refuse to abandon—those become the architecture of what comes next. This is the great leap forward. Not into technology. Into ownership of a future that's being constructed in real time. And history will remember who showed up when the systems were still being written.

"You don't need permission to continue. You're allowed to be powerful here. You're allowed to be wealthy here. You're allowed to be visible here. You're allowed to be fully human here."

ABOUT THE AUTHOR

Erica Rooney is a leadership architect, founder, and nationally recognized voice at the intersection of women, power, and artificial intelligence. She is the founder of HER Collective, a fast-growing leadership community for ambitious women in corporate environments who are ready to expand their confidence, strategic influence, and economic impact. Through HER Collective, Rooney has built more than a network. She has created a room where women rise together, lift while they climb, and actively reshape the systems they lead within.

Rooney is the author of *Glass Ceilings and Sticky Floors* and the host of the top-rated podcast *Glass Ceilings and Sticky Floors*, where she interviews bold thinkers and executives redefining leadership in the modern workplace. She is also the co-host of *AI Voice or Victim*, a forward-looking podcast examining how artificial intelligence is transforming careers, and how women can claim power in that transformation rather than be displaced by it.

With more than fifteen years of executive leadership experience, Rooney has advised organizations across industries on talent strategy, culture, and future-of-work readiness. Often described as a translator between tech and trust, she helps women build fluency in AI while strengthening the confidence required to wield it strategically.

Her commitment to access extends beyond corporate leadership. Rooney gives back through nonprofit partnerships, including Women Applying AI, The American Heart Association, and Dress for Success, advancing economic mobility and technological literacy for women at every stage of their careers.

Based in North Carolina, Rooney is building a legacy anchored in one belief: **when women gather with ambition, clarity, and conviction, they do not simply succeed, they shift power.**

ACKNOWLEDGMENTS

This book was not written alone.

Dan, thank you for being the steady ground beneath every big leap. For believing in my ambition even when it stretched us. For holding the center of our home while I chased ideas that did not yet have names. Hudson and Halle, you are my why. Watching you grow reminds me exactly why this work matters, because the future is not abstract. It is sitting at our dinner table.

To the women of HER Collective—you are the heartbeat of this movement. You said yes before there was proof. You joined before there was polish. You trusted the vision when it was still just a sketch in a notebook. Caitlin Moi, Ashley Tryban, and Molly Kirk—you leaned all the way in. You stayed for year two. You shaped what HER is today. Your belief in me, and in what we are building together, changed everything. This book carries your fingerprints.

Greg Boone, my business partner and co-host—thank you for seeing something in me before I saw it in myself. You were the first to name what I was doing as technical, strategic, and AI-forward. For someone who never identified as "technical," that shift mattered more than you know. You helped me claim a part of my voice that is now central to this work.

Ally Berthiaume, thank you for sharpening these pages with both rigor and care for not just one book, but two. And thank you to Paul Baillie-Lane at WorldChangers Media for recognizing the urgency of this message and helping bring it into the world.

And finally, to you—the woman holding this book. Whether you are confident in your place in the future of work or still questioning if you belong, you are part of this story now. This movement grows because you are here.

We lift while we climb.

ENDNOTES

Introduction

1 Pew Research Center. “Gender Pay Gap in U.S. Has Narrowed Slightly Over 2 Decades.” March 4, 2025. https://www.pewresearch.org/short-reads/2025/03/04/gender-pay-gap-in-us-has-narrowed-slightly-over-2-decades/.

2 Hinchliffe, Emma. “Women CEOs Run 11% of Fortune 500 Companies in 2025.” Fortune, June 2, 2025.

3 Christensen, MacKenzie A. “Tracing the Gender Confidence Gap in Computing: A Cross-National Meta-Analysis of Gender Differences in Self-Assessed Technological Ability.” Social Science Research 111 (March 2023): 102853.https://doi.org/10.1016/j.ssresearch.2023.102853.

4 McKinsey & Company and LeanIn.Org, Women in the Workplace 2023 (New York: McKinsey & Company, 2023), https://www.mckinsey.com/featured-insights/diversity-and-inclusion/women-in-the-workplace.

5 Konrad, Alison M. “Women Are Avoiding Using Artificial Intelligence. Can That Hurt Their Careers?” Harvard Business School Working Knowledge (Baker Library | Bloomberg Center), November 20, 2023. https://www.library.hbs.edu/working-knowledge/women-are-avoiding-using-artificial-intelligence-can-that-hurt-their-careers.

6 McKinsey and Lean In Women in the Workplace 2023.

7 Koprek, Julia. "Women in Tech Statistics: 2023 Numbers That Reveal the Tech Gender Gap." Lemon.io Blog, May 20, 2023.

8 https://openai.com/index/openai-announces-new-members-to-board-of-directors/.

9 "Lila Ibrahim," WomenTech Network, accessed October 21, 2025, https://www.womentech.net/women-in-tech/Lila/Ibrahim/128061.

10 "The Inspiring Story of Daniela Amodei, Anthropic's Leader," KitRUM, May 2, 2024, https://kitrum.com/blog/the-inspiring-story-of-daniela-amodei-anthropics-leader/.

11 Brooks, Alison Wood. "Breaking Through the Self-Doubt That Keeps Talented Women from Leading." August 22, 2023.

12 Cortland, Clarissa, Micah A. E. Schmitz, and Amanda B. Diekman. "The Gender Stereotyping of Artificial Intelligence." European Journal of Social Psychology 53, no. 2 (2023).

13 Angelucci, Manuela, Brian G. Knight, and Na'ama Shenhav. "AI Hiring Tools Exhibit Complex Gender and Racial Biases." VoxDev, July 24, 2023.

14 U.S. Bureau of Labor Statistics. "Usual Weekly Earnings of Wage and Salary Workers, Fourth Quarter 2024." January 22, 2025.

15 https://www.library.hbs.edu/working-knowledge/women-are-avoiding-using-artificial-intelligence-can-that-hurt-their-careers?

16 Northeastern University, "Gender Differences in AI Risk Perception," Northeastern Global News, January 22, 2026, https://news.northeastern.edu/2026/01/22/gender-differences-ai-risk-research/.

Chapter 1

17 Al Hamid, Abdullah, et al. "Gender Bias in Diagnosis…" Cureus 16, no. 2 (2024).

18 Consumer Reports. "A Crash Test Bias Puts Female Drivers at Risk." February 2023.

19 Goldin, Claudia, and Lawrence F. Katz. The Race Between Education and Technology. Cambridge: Harvard University Press, 1998.

20 Tatman, Rachael. "Gender and Dialect Bias in YouTube's Automatic Captions." In Proceedings of the First ACL Workshop on Ethics in Natural Language Processing, 53–59. Valencia: Association for Computational Linguistics, 2017. https://aclanthology.org/W17-1606.

21 David 1990.

22 David 1990; Goldin and Katz 1998.

23 David 1990.

24 Kline, Patrick, and Enrico Moretti. "Local Economic Development…" Quarterly Journal of Economics 129, no. 1 (2014).

25 Guider. "How Mentoring Empowers Women in Their Careers." Accessed 2025.

26 Bridle, James. "Ada Lovelace, the First Tech Visionary." The New Yorker, October 30, 2015.

27 Kay, Katty, and Claire Shipman. The Confidence Code. New York: HarperBusiness, 2014.

28 Kay and Shipman 2014.

29 Konrad, Alison M. "Women Are Avoiding Using Artificial Intelligence. Can That Hurt Their Careers?" Harvard Business School Working Knowledge (Baker Library | Bloomberg Center), November 20, 2023. https://www.library.hbs.edu/working-knowledge/women-are-avoiding-using-artificial-intelligence-can-that-hurt-their-careers.

30 Konrad and Alison 2023.

31 Deloitte. "Women and Generative AI." 2024.

32 Deloitte 2024.

33 Pew Research Center. 2025 AI Survey.

34 Riva, Elena, and Maddalena Marini. 2025.

35 Elena and Marini 2025.

36 Deloitte 2024.

37 Konrad and Alison 2023.

38 Konrad and Alison 2023.

39 Konrad and Alison 2023.

40 Kray, Laura J., Lauren C. Howe, and Hillary Anger Elfenbein. "Trust the Machine or Trust Yourself?" Current Opinion in Psychology 54 (2024).

Chapter 2

41 Dastin, Jeffrey. "Amazon Scraps Secret AI Recruiting Tool That Showed Bias Against Women." Reuters, October 10, 2018.

42 O'Malley, Michael. Keeping Watch: A History of American Time. Washington, DC: Smithsonian Institution Press, 1990.

43 Hasyim, Said. "The Role of Visualization in Achieving Goals." July 9, 2023.

44 Locke, Edwin A., and Gary P. Latham. "Building a Practically Useful Theory of Goal Setting." American Psychologist 57, no. 9 (2002).

45 Lagorio-Chafkin, Christine. "Bumble Says Its AI Can Detect Lewd Photos." Inc., August 5, 2021.

46 Lagorio-Chafkin 2021.

47 Dockterman, Eliana. "Whitney Wolfe Herd's Dating App Empire." TIME, March 4, 2021.

48 Hart, Jordan. "Bumble CEO Says AI Can Help Revolutionize Dating." Business Insider, September 13, 2023.

49 Hart 2023.

50 Hasyim 2023.

Chapter 3

51 U.S. Census Bureau. Educational Attainment in the United States: 2024.

52 McKinsey & LeanIn.Org. Women in the Workplace 2023.

53 https://www.google.com/url?q=https://www.weforum.org/stories/2024/03/empowering-women-in-stem-how-we-break-barriers-from-classroom-to-c-suite/.

54 Chan, Joseph B., and Matt Sigelman. "Why Internal Mobility Is So Important—and How to Get It Right." Harvard Business Review, May 5, 2021. https://hbr.org/2021/05/why-internal-mobility-is-so-important-and-how-to-get-it-right.

55 Techopedia. "The Gender Gap in Tech 2022."

56 Northcote, Stafford H., and C. E. Trevelyan. Report on the Organisation of the Permanent Civil Service. London, 1854.

57 An, Jiafu, Difang Huang, Chen Lin, and Mingzhu Tai. "AI Hiring Tools Exhibit Complex Gender and Racial Biases." 2023.

58 McKinsey and LeanIn.Org. Women in the Workplace 2023.

59 https://www.weforum.org/stories/2025/01/future-of-jobs-report-2025-jobs-of-the-future-and-the-skills-you-need-to-get-them.

60 Xu, Jing. "The Gender Gap in Executive Promotions." Journal of Corporate Finance 89 (2024).

61 McKinsey and Lean In 2023.

62 MIT Sloan. "Women Are Less Likely Than Men to Be Promoted." March 31, 2022.

63 Exley, Christine L., and Judd B. Kessler. "Why Don't Women Self-Promote As Much As Men?" 2019.

64 Chand, Suma P., Daniel P. Kuckel, and Martin R. Huecker. "Cognitive Behavior Therapy." 2025.

65 McKinsey and Lean In 2023.

66 McKinsey and Lean In 2023.

67 https://www.google.com/url?q=https://hbr.org/2021/02/

promotions-arent-just-about-your-skills-theyre-about-your-relationships?utm_source%3Dchatgpt.com&sa=D&source=docs&ust=1774524999134491&usg=AOvVaw1PLC6Bc-DOHi6_zSH84JT-Z.

68 Techopedia 2022.

69 Bowles, Hannah Riley, Linda Babcock, and Lei Lai. "Social Incentives for Gender Differences in Negotiations." 2007.

70 Fragale, Alison. Likeable Badass. 2024.

71 Fragale 2024.

72 Fragale 2024.

Chapter 4

73 Lauer, Josh. Creditworthy. 2017.

74 Osman, Maddy. "Mind-Blowing LinkedIn Statistics." 2025.

75 Osman 2025.

76 Osman 2025.

77 Ibarra, Herminia, Nancy M. Carter, and Christine Silva. "Why Men Still Get More Promotions Than Women." Harvard Business Review, 2010.

78 Arruda, William. "How Women Can Use Personal Branding to Accelerate Career Success." Forbes, 2024.

79 Apollo Technical. "Social Media Recruiting Statistics." 2025.

80 Byezhanova 2025.

81 Reynolds, Justin. "AI for Thought Leadership." 2025.

82 Byezhanova, Marina. "Why Thought Leadership Wins in the AI Era." Inc., 2025.

83 Byezhanova 2025.

84 Oladipo, Tamilore. "No-Pressure System That Helped Me Stay Consistent." 2025.

85 Timm, Tricia Montalvo. "The Perfectionism Pitfall." 2023.

Chapter 5

86 Reill, Amanda. "A Simple Way to Make Better Decisions." 2023.
87 Reill 2023.
88 Gamache, Daniel L., et al. "Shifting Perspectives." 2023.
89 Landau, C. "The Third Shift." 2020.
90 Landau 2020.
91 Baron, Jonathan, and John C. Hershey. "Outcome Bias in Decision Evaluation." 1988.
92 Heilman, Madeline E., and Tyler G. Okimoto (2007). "Why Are Women Penalized for Success at Male Tasks? The Implied Communality Deficit." Journal of Applied Psychology, 92(1), 81–92.
93 Westover, Jonathan H. "The Total Cost of Difficult Leaders." 2025.
94 De Smet, Aaron; Bonnie Dowling; Bryan Hancock; and Bill Schaninger. "The Great Attrition Is Making Hiring Harder. Are You Searching the Right Talent Pools?" McKinsey & Company, July 13, 2022. Accessed November 7, 2025. https://www.mckinsey.com/capabilities/people-and-organizational-performance/our-insights/the-great-attrition-is-making-hiring-harder-are-you-searching-the-right-talent-pools?cid=other-soc----oth---&sid=7784742232&linkId=184202328.
95 Larson, Erik. "49% of Companies Have Stagnant Decision Styles." 2024.
96 Josie Cox, "Over the Last Two Decades, the U.S. Gender Pay Gap Has Hardly Moved," Forbes, March 5, 2025, https://www.forbes.com/sites/josiecox/2025/03/05/gender-pay-gap-has-hardly-moved/.
97 Porter, Theodore M. The Rise of Statistical Thinking. 1986.
98 Wedell-Wedellsborg, Thomas. "Do You Understand the Problem You're Trying to Solve?" HBR On Leadership (podcast), April 3, 2024. Accessed November 7, 2025. https://hbr.org/podcast/2024/04/do-you-understand-the-problem-youre-trying-to-solve.

99 Huston, Therese, Ph.D. "Make Better Decisions." HBR IdeaCast (podcast episode 524), May 12, 2016. Accessed November 7, 2025.

100 Powell, Melissa. "Are Cognitive Biases Impacting Your Decision-Making?" Chief Executive, August 29, 2024. Accessed November 7, 2025. https://www.chiefexecutive.net/are-cognitive-biases-impacting-your-decision-making/.

101 Larson 2024.

Chapter 6

102 Hounshell, David A. From the American System to Mass Production. 1984.

103 Konrad 2023.

104 Singh, Ben, et al. "Time to Form a Habit." 2024.

105 Risko, Evan F., and Sam J. Gilbert. "Cognitive Offloading." 2016.

106 World Health Organization. "Burn-out an 'Occupational Phenomenon': International Classification of Diseases." World Health Organization, May 28, 2019. Accessed December 1, 2025 https://www.who.int/news/item/28-05-2019-burn-out-an-occupational-phenomenon-international-classification-of-diseases.

107 World Health Organization 2019.

Chapter 7

108 Schaeffer, Katherine. Pew Research Center, 2025.

109 National Women's Law Center. "The Wage Gap Robs Mothers." 2025.

110 Felix Salmon, "Women MBAs Are Asking for Higher Pay, But Still Earning Less," Axios, May 15, 2024, https://www.axios.com/2024/05/15/women-salary-negotiate-gender-pay-gap?utm_source=chatgpt.com.

111 Cortes, Patricia, et al. NBER Working Paper No. 32154, 2024.

112 Bowles, Hannah Riley, Linda Babcock, and Lei Lai. "Social Incentives for Gender Differences in the Propensity to Initiate Negotiations: Sometimes It Does Hurt to Ask." Organizational Behavior and Human Decision Processes 103, no. 1 (January 2007): 84–103. https://hbr.org/2014/04/why-women-dont-negotiate-their-job-offers.

113 World Economic Forum. Global Gender Gap Report 2024.

114 Kelly Cardin and Thelma Akpan, "The Pay Transparency Laws to Know in 2025 in the United States and Beyond," Littler, February 21, 2025, https://www.littler.com/news-analysis/asap/pay-transparency-laws-know-2025-united-states-and-beyond?utm_source=-chatgpt.com.

115 Taylor, Frederick Winslow. The Principles of Scientific Management. New York: Harper & Brothers, 1911.

116 World Economic Forum 2024.

117 Bowles, Hannah Riley and McGinn, Kathleen L. (2008). "Gender in Job Negotiations: A Two-Level Game". Negotiation Journal, 24(4), 393–410.

118 Cortes et al 2024.

119 Bowles and McGinn 2008.

120 Jenny Donovan, "How Organizations Can Take a Systematic Approach to Pay Equity," SHRM Executive Network Insights, March 11, 2024, https://www.shrm.org/executive-network/insights/systematic-approach-pay-equity?utm_source=chatgpt.com. shrm.org.

Chapter 8

121 Board of Governors of the Federal Reserve System. Survey of Consumer Finances, 2022.

122 Fry, Richard. Pew Research Center, 2024.

123 Fry 2024.

124 Fry 2024.

125 National Women's Law Center. "The Wage Gap for Mothers and Fathers". Washington, DC: National Women's Law Center, updated 2023–2024.

126 Vanguard. How America Saves 2025.

127 Cerulli Associates. The Cerulli Report. 2022.

128 McKinsey & Company. Women as the Next Wave of Growth in Wealth Management. 2020.

129 Fry 2024.

130 Konrad 2023.

131 U.S. Bureau of Labor Statistics. Highlights of Women's Earnings in 2023. February 2024. https://www.bls.gov/opub/reports/womens-earnings/2023.htm.

132 U.S. Bureau of Labor Statistics 2024.

133 McKinsey 2020.

134 Vanguard 2025.

135 Fidelity Investments. 2024.

136 Fidelity 2024.

137 OECD. Algorithm and Eve. 2024.

138 U.S. Securities and Exchange Commission. "Investment Adviser Public Disclosure."

139 Consumer Financial Protection Bureau. "Bringing Dark Patterns to Light". Washington, DC: Consumer Financial Protection Bureau, 2022.

Chapter 9

140 World Economic Forum. Skills Outlook 2025.

141 Brynjolfsson, Erik, Danielle Li, and Lindsey Raymond. 2023.

142 Jack Zenger and Joseph Folkman, "Research: Women Are Better Leaders During a Crisis," Harvard Business Review, December 30, 2020.

143 Zenger and Folkman 2020.

144 Susanne Hupfer, Bree Matheson, Gillian Crossan, Ariane Bucaille, and Jeff Loucks, "Women and Generative AI: The Adoption Gap Is Closing Fast, but a Trust Gap Persists," Deloitte Insights, November 19, 2024 https://www.deloitte.com/us/en/insights/industry/technology/technology-media-and-telecom-predictions/2025/women-and-generative-ai.html.

145 Woolley, Anita Williams, Thomas W. Malone, Christoph Riedl, and Nathaniel L. Foote. "Authentic Leadership and Employee Flourishing: A Multi-Sector Analysis." Frontiers in Psychology 11 (2020). doi:10.3389/fpsyg.2020.01572.

146 Alilyyani, B., Carol Wong, and Greta Cummings. "Antecedents, Mediators, and Outcomes of Authentic Leadership in Healthcare: A Systematic Review." International Journal of Nursing Studies 83 (2018): 34–64.

Chapter 10

147 Chawla, Nitesh V., Yang Yang, Brian Uzzi, and Richard L. Thomas. "A Network's Gender Composition and Communication Pattern Predict Women's Leadership Success." Proceedings of the National Academy of Sciences of the United States of America 116, no. 28 (2019): 13850-55. https://doi.org/10.1073/pnas.1721438116.

148 Carr, David, Jennifer C. Lerner, Baiyun Gong, and John Keating. "The Value of Belonging at Work." Harvard Business Review, December 2019. https://hbr.org/2019/12/the-value-of-belonging-at-work.

Chapter 11

149 World Economic Forum. Global Gender Gap Report 2024.

150 World Economic Forum. Future of Jobs Report 2025.

151 OECD. Algorithm and Eve. 2024.

152 Swaminathan, Viren, Hannes Schwandt, Christopher E. Choi, and David Blumenthal. "Gender Differences in the Effect of Physician Gender on Mortality among Patients with Acute Myocardial Infarction." Proceedings of the National Academy of Sciences 115, no. 34 (2018): 8569–8574. https://doi.org/10.1073/pnas.1800097115.

153 Rogers, Cody J., and Chris O. Cook. "Female Crash Test Dummy Safety and Injury Risk." University of Virginia Center for Applied Biomechanics, 2019.

154 OECD 2024.

155 Gmyrek, Paweł, Janine Berg, and David Bescond. Generative AI and Jobs: A Global Analysis of Potential Effects on Job Quantity and Quality. ILO Working Paper No. 96. Geneva: International Labour Organization, 2023.

156 Deloitte 2024.

157 Brynjolfsson, Li, and Raymond 2025.

WORLDCHANGERS MEDIA